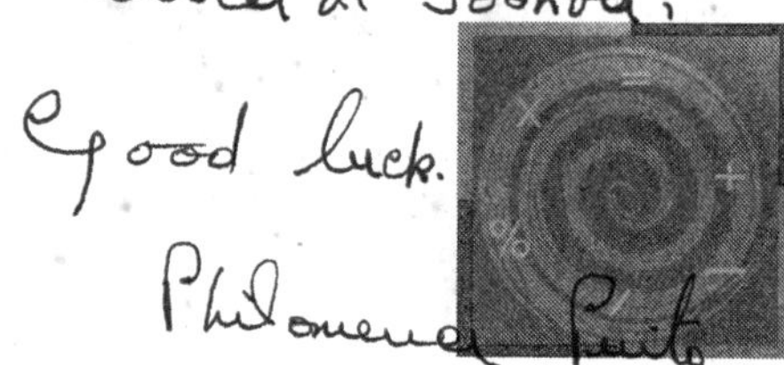

VEDIC MATHS 2

Vedic Formulae *(Sutras)* Explained in Simple Methods

H. K. Gupta

BPI INDIA PVT LTD

BPI INDIA PVT LTD

Vedic Maths 2

H.K.Gupta

ISBN 81-7693-283-3
First Print 2006

For Information Contact:
BPI INDIA PVT LTD
16, Darya Ganj, Next to Castle Guest House,
New Delhi - 110 002
Tel.: 2328 4898, 2327 6118, Fax: 2327 1653
E-mail: bpipl@vsnl.com, bpiindia@airtel-broadband.com

Concept, design and artworks by Neve Designs
Tel.:4166 5006, 9810655355
E-mail: nevedesigns@airtel-broadband.com

Printed in India by Gopsons Papers Ltd., Noida

About the Author

From an early age, H.K. Gupta displayed an unmistakable flair for mathematics. This gift made itself increasingly apparent as he pursued his higher studies, and it came as no surprise when he was selected for the prestigious IIT, Bombay, from where he earned his Bachelor`s degree in Electrical Engineering.

Throughout his meritorious career in the Indian Air Force – he left when he was a Wing Commander – Gupta nourished his keen interest in the subject by teaching school children both maths and physics, and coaching aspirants for appearing in All India competitive examinations—activities that still mean a lot to him.

A chance meeting in 2001 inspired him to explore for himself the subtle rhythms and intricate formulae embedded in the timeless Vedas—a wonder-world whose arcane secrets he gradually unraveled. This fascinating and mind-expanding book is the result of his plunge into the little-known world of *Vedic Mathematics*.

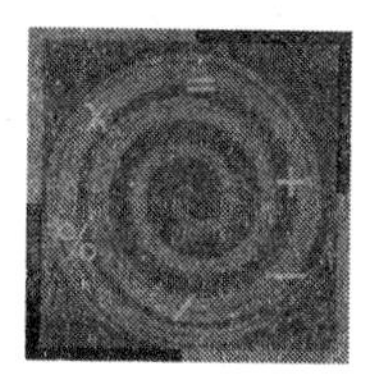

Contents

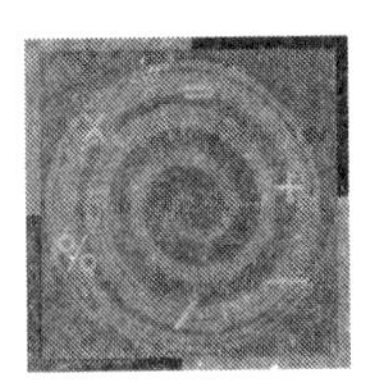

Introduction

The first book on vedic maths deals with simple arithmetical calculations like multiplication, division and finding the square as well as square roots of the numbers. In continuation, the present book deals with more complicated arithmetical calculations like finding the cube and cube roots of the numbers, decimal values of fractions as well as the application of vedic methods in higher maths.

It must be emphasized that vedic maths not only provides intitutive short cuts but also different methods for solving the same problem and thus makes learning maths more of a pleasure.

For instance 16^3 can be found in three ways.

First method: Anurupya Sutra (Refer to chapter 1 for explanation)

$16^3 =$	1	6	36	216	
		12	72		
	3	12	21		
	4	0	9	6	Ans

Second method: Yavadunam Sutra (for cubing)

16^3 = 28/ 108/ 216
= 4096 Ans

Third method: Binomial Expression.

Similar to the first method but methodology is different.

Thus the student gets the choice of using any of the above methods.

After introducing zero notation the vedic system has been advocating decimal system not because of any partiality but solely on its intrinsic merits. For instance, there is only one method of calculating the decimal value of the fraction 1/19 in conventional method whereas the vedic method gives a student two alternatives to choose from. Both these methods are much superior and labour saving as compared to the conventional method.

Thus 1/19 by conventional method

```
19) 1·00(·052631578947368 21
      95
      50
      38
      120
      114
        60
        57
         30
         19
         110
          95
          150
          133
            170
             152
               180
                171
                   90
                   76
```

140
133
70
57
130
114
160
152
80
76
40
38
20
19
1

The above method requires 18 long steps of division before its recurring decimal value

1/19= ·052631578947368421 is obtained. However, there are two ways in Vedic mathematics to find the decimal value of 1/19.

First vedic method:

Starting from the last digit and making use of Ekadhika Purva Sutra we can write:

1/19= ·052631578/947368421 (Chapter 3)

Second vedic method:

In this we use make use of the Auxiliary fraction and write:

1/19= = ·0 5 2 6 3 1 5 7 8 9 4 7 3 6 8 2 1
1 0 1 0 0 1 1 1 1 0 1 0 1 1 0 0 0 (Chapter 4)

Notice that in the first method we write the last i.e. right hand most digit and then move towards the left whereas in the second method we start from the left and move to the right.

This gives a student a choice and makes learning maths fun.

Finding Cubes of Numbers

First method: The Anurupya Sutra

This method presupposes prior knowledge of cubes of single digit numbers from 1 to 9. To find out cubes of two digit numbers, known identity of Algebra is used.

$$(a + b)^3 = a^3 + 3a^2b + 3ab^2 + b^3.$$

Almost every student of mathematics knows this but very few make use of it. The above identity can be written as

$$\begin{array}{l} (a + b)^3 = a^3 + a^2b + ab^2 + b^3 \\ \qquad\qquad\qquad\quad 2a^2b \quad 2ab^2 \end{array}$$

We have broken $3a^2b$ and $3ab^2$ into two parts $3a^2b$ as $a^2b + 2a^2b$ and $3ab^2$ as $ab^2 + 2ab^2$ to simplify the calculations. If one scrutinizes the four terms placed at the top one can easily notice that they are in geometrical progression (G.P) with the common ratio being b/a. This is a significant finding which will help us in calculating the cube of two-digit numbers.

The above method is further explained with the help of a few examples.

Example 1: Find 16^3

a b

Here a=1, b=6 and b/a =6.

Step 1. Find $a^3 = 1^3 = 1$.

Step 2. Find $a^2b = a^3b \times b/a = 1 \times 6 = 6$.

Step 3. Find $ab^2 = a^2b \times b/a = 6 \times 6 = 36$.
Step 4. Find $b^3 = ab^2 \times b/a = 36 \times 6 = 216$.

All the calculations are completed and the answer is obtained as shown below.

$16^3=$	1	6	36	216	
		12	72		
	<u>3</u>	<u>12</u>	<u>21</u>		Remainder at each stage.
	4	0	9	6	Answer

When the number in any of the above calculations contains more than one digit, the right hand most digit thereof is to be taken and preceding left hand digits or digit should be carried over to the left hand placed under the previous digit or digits of the upper row. The digit carried over may be shown in the working. Thus, in the above example we take right hand most digit 6 of the number 216 as the right hand most digit of the answer and carry over 21 and place it below the upper rows. Similarly, the right hand digit of the summation (36 + 72 + 21 = 129), 9, is the second right hand digit of the answer and 12 is carried over and placed below the upper rows. Next, the right hand digit of the summation (6 + 12 + 12 = 30), 0 is the third digit of the answer and 3 is carried over and placed below the upper row. Finally (1 + 3) = 4 is the last digit of the answer.

Example 2: Find 23^3

Here a=2, b=3 and b/a =3/2.

Step 1. Find $a^3 = 2^3 = 8$.
Step 2. Find $a^2b = a^3 \times b/a = 12$.
Step 3. Find $ab^2 = a^2b \times b/a = 12 \times 3/2 = 18$.
Step 4. Find $b^3 = ab^2 \times b/a = 18 \times 3/2 = 27$.

Answer can be written as shown.

$23^3=$	8	12	18	27	
		24	36		
	4	5	2----------		Remainder at each stage.
	12	1	6	7	Answer

Example 3: Find 97^3

Here a=9, b=7 and b/a= 7/9.

Step 1. Find $a^3 = 9^3 = 729$.

Step 2. Find $a^2b = a^3 \times b/a = 567$.

Step 3. Find $ab^2 = a^2b \times b/a = 441$.

Step 4. Find $b^3 = ab^2 \times b/a = 343$.

Now the answer can be written as shown.

$97^3=$	729	567	441	343	
		1134	882		
	183	135	34————		Remainder at each stage.
	912	6	7	3	Answer

97^3 can also be found very easily by putting

$97^3 = (100-3)^3$ here a=100, b=-3 and making use of identity

$(a - b)^3 = a^3 - a^2b + ab^2 - b^3$

$\qquad - 2a^2b + 2ab^2$

$(100-3)^3 = 100000 - 30000 + 900 - 27$

$\qquad -60000 + 1800$

$= 100000 - 9000 + 2700 - 27$

$= 912673$ Answer

To ease the calculations, in the above three examples, we have calculated b/a that is not really necessary. It is sufficient to identify a, b and do the rest of the calculations mentally as shown below.

Example 4: Find 11^3

$11^3=$	1	1	1	1	
		2	2		
	1	3	3	1	Answer

Example 5: Find 12^3

$12^3 =$	1	2	4	8	
		4	8		
	1	7	2	3	Answer

Example 6: Find 8^3

$8^3 = (10-2)^3$

$= 1000-200+40-8$

$-400+80$

5	1——		Remainder
5	1	2	Answer

Example 7: Find 24^3

$24^3 =$	8	16	32	64	
		32	64		
	5	10	16———		Remainder
	13	8	2	4	Answer

Example 8: Find 32^3

$32^3 =$	27	18	12	8	
		36	24		
	5	3—————			Remainder
	32	7	6	8	Answer

Example 9: Find 19^3

$19^3 =$	1	9	81	729	
		18	162		
	5	31	72----------		Remainder
	6	8	5	9	Answer

Example 10: Find 17^3

$17^3 =$	1	7	49	343	
		12	98		
	3	18	34-----------		Remainder
	4	9	1	3	Answer

Example 11: Find 62^3

$62^3=$	216	72	24	8	
		144	48		
	22	7------------------			Remainder
	238	3	2	8	Answer

Compare this with the actual multiplication required to obtain 62^3.

$62^3 = 62 \times 62 \times 62$

```
62 x 62=  124
          372
         3844
          x62
         7688
       23064
       238328            Answer
```

Obviously Vedic method is much shorter.

Second method: The Yavdunam Sutra

We have already applied the magical Yavdunam Sutra in finding squares of numbers (Refer to chapter 5 page 53 of book 1).

The same sutra can easily be applied in finding cube of a number with certain modifications. Suppose we wish to ascertain the cube of 105. Our base being 100 the excess is 5. So we add not 5 as in case of squaring operation but its double that is 2x5=10 and thus have 105+10=115 as the left hand most portion of the cube.

Next we put down the new excess multiplied by the original excess that is 15x5=75 as the middle portion of the answer. Finally we affix the cube of the original excess that is $5^3=125$ as the last portion of the answer. Now the answer is complete and we can write

$105^3 = 115/75/125$

1-----	Remainder
1157625	Answer

A few more examples are given to illustrate the above method.

Example1: Find 112^3

Step 1. In 112, our base number is 100 and we have 12 in excess. Thus the left-hand digits are 112+12x2=136.

Step 2. The middle portion of the answer is 36x12=432.

Step 3. The right hand portion is 12^3= 1728.

The answer now can be written as:

112^3= 136/432/1728

4 17-------	Remainder
1404928	Answer

Example2: Find 103^3

Step 1. In 103, 3 is excess of the base number 100. Thus the left-hand digits are 103+3x2=109.

Step 2. The middle portion is 9x3=27.

Step 3. The right hand portion is 3^3=27.

The answer now can be written as:

103^3= 109/27/27

1092727 Answer

Example3: Find 93^3

Step 1. In 93. 7 is less than the base number 100. Thus the left-hand digits are 93-7x2=79.

Step 2. The middle portion is (-21)x(-7)=147.

Step 3. The right hand portion is $(-7)^3$= -343.

The answer now can be written as:

$93^3 =$ 79/147/-343
1 -4-------- Remainder
804357 Answer

As the last digit is negative, we subtact it from 4000. Thus 400-343=57 and carry -4 to the next stage.

Example 4: Find 113^3

113^3=139/507/2197 (because 39x13=507 and 13^3=2197)

139/07/97
5 21-------- Remainder
1442897 Answer

Example 5: Find 1004^3

1004^3=1012/048/064 (because 12x4=48 and 4^3=64)

=1012048064 Answer

Example 6: Find 996^3

996^3=988/(-12x-4)/-4^3

988/048/-064

=98804736 Answer

Example 7: Find 1007^3

1007^3=10021/0147/0343

=1002101470343 Answer

Remember, the number of digits from the last number to be taken in the final answer must be one less than the number of digits in the number whose cube is being calculated. Thus, while finding cube of 105 the last number is 5^3 =125, here we take two digits i.e., 25 which is one digit less than 3 digits contained in 105. Similarly, while finding cube of 1004, a number containing 4 digits, the last number is 4 x 4 x 4 = 64. But in final answer we must have (4 –1 =3) digits. Thus we make it a three digit number by preceding it with a zero and making it 064.

Exercise 1

a) A cubical container has its side =14m find its volume.

b) Find 18^3.

c) Find 46^3.

d) Calculate Cube of following numbers:

i) 26^3 ii) 49^3 iii) 62^3 iv) 55^3 v) 59^3

vi) 47^3 vii) 33^3 viii) 29^3 ix) 1007^3 x) 103^3

xi) 1005^3

Cube Root of Numbers

Consider cubes of single digit numbers.

	Last digit
$1^3=1$	1
$2^3=8$	8
$3^3=27$	7
$4^3=64$	4
$5^3=125$	5
$6^3=216$	6
$7^3=343$	3
$8^3=512$	2
$9^3=729$	9

From the above it is clear that

i) They all have their own distinct endings and there is no possibility of overlapping or doubt as in case of squares.

ii) The last digit of 2^3 is 8 and 3^3 is 7 and last digit of $8^3=512$ is 2 and that of $7^3=343$ is 3.

We shall utilize the above information in finding out cube root of exact cubes as explained below:

Step 1: Start from the right hand side of the number. When the three digits are over, put a comma. Thus what we are doing is grouping the numbers into three digit numbers including a single or double digit if any. For example:

i) 9, 261
ii) 226, 981
iii) 105, 823, 817
iv) 33, 076, 161
v) 62, 741, 116, 007, 421
vi) 91, 010, 000, 000, 468
vii) 1, 548, 816, 893
viii) 73, 451, 930, 798

Step 2: Number of digits in the cube root will be equal to the number of groups in the original number. Thus the number of digits in cube root in the above examples will be:

i) 2
ii) 2
iii) 3
iv) 3
v) 5
vi) 5
vii) 4
viii) 4

Step 3: The first group of the original numbers decides the first digit of the cube root. Thus the first digit of the cube root on mere inspection in above examples is:

i) 2
ii) 6
iii) 4
iv) 3
v) 3
vi) 4
vii) 1
viii) 4

Step 4: The last digit of the number decides the last digit of the cube root. Here our knowledge of the cubes of single digit numbers helps us. Thus the last digit of cube root in above examples will be:

i) 1
ii) 1
iii) 3
iv) 1
v) 1
vi) 2
vii) 7
viii) 2

Thus before proceeding further we already have information regarding the number of digits in a cube root, the first digit and the last digit by the well known Vedic principle of Vilokanam (inspection). Thus by mere inspection of following numbers we can write:

	Number	Number of digit in the cube root	First digit	Last digit
i)	9,261	2	2	1
ii)	226,981	2	6	1
iii)	105,823,817	3	4	3
iv)	33,076,161	3	3	1
v)	62,741,116,007,421	5	3	1
vi)	91,010,000,000,468	5	4	2
vii)	1,548,816,893	4	1	7
vii)	73,451,930,798	4	4	2

Having found out the number of digits, the first digit and the last digit we adopt the same procedure as in the case of square roots to find out the other digits of the cube root. The only difference is that our divisor in this context will not be double the first digit of the root but three times the square there of.

As we know the first digit at the very outset our chart will look like as shown in the following example.

Example 1: Find the cube root of 13824

Step 1. The inspection tells that the number of digits in the cube root will be 2.

Step 2. First digit of the cube root=2.

Step 3. Now let us find out whether this is a perfect cube and the other digits of the answer.

13824=13: 8 2 4

 12   5 10 6

 2 4 0

Step 4. Divisor=$3a^3$=3x4=12 and remainder ($13-2^3$)=5 is written as shown.

Step 5. We divide 58 with 12 without subtracting anything. Write Q1=4, remainder R1=10 as shown.

Step 6. Our gross dividend is $102-3ab^2$ i.e. 102-3x2x16=6.

Step 7. Next we divide 6 with 12 and write Q2=0, R2=6.

Step 8. Now our gross dividend is $64-(6abc+b^3)$ i.e. 64-(6x2x4x0+64)=0.

So the cube root of 13824 = 24. It is a perfect cube.

The other method available is to find out factors of 13824 as shown below which is a time consuming and cumbersome method.

2) 13824
2) 6912
2) 3456
2) 1728
2) 864
2) 432

2) 216
2) 108
2) 54
3) 27
3) 9
3) 3
0

13824=2x2x2x2x2x2x2x2x2x3x3x3

Therefore cube root = 2x2x2x3
= 24

The technique outlined in the above example is valid for finding out cube roots in general. It must be emphasized that this is a good method of finding approximations. Like in the above example mere inspection would have given cube root of 13824=24.

This will be further confirmed by:

24^3=8	16	32	64	
	32	64		
	10	6-----------		Remainder
13	8	2	4	

Consider another example:

Example 2: Find the cube root of 258474853

258474853=258: 4 7 4 8 53
108 42 100 89 111 47
6 37 0

Step 1. Mere Viloknam (inspection) tells us that the cube root will have 3 digits.

Step 2. The first digit of the cube root is 6 as 6^3=216 is less than 258 and 7^3= 353 is greater than 258 and therefore inadmissible.

Step 3. The first Q and R are therefore 6 and 258-216=32 respectively. They are put down as shown above.

Step 4. The second gross dividend is 424. Nothing is to be subtracted from this. It is to be divided by 108. Q1=3 and R1=100 are written down in appropriate places as shown above.

Step 5. Now the gross dividend is 1007. Subtract $3ab^2$ i.e. $3 \times 6 \times 3^2=162$ from it. Working dividend=1007-162=845.

Dividing by 108 gives us Q2=7 and R2=89 and are written down in proper places.

Step 6. Next gross dividend is 894. Subtract from this

$6abc+b^3=6 \times 6 \times 3 \times 7+27$

$=783$

So the working dividend 894-783=111. Divide this by 108. The Q3=1 and R3=3 but if we take the next dividend as 36 and subtract $3ab^2+3b^2c$ i.e, 882+189. We shall land up with a negative number. Therefore we write Q3=0 and R3=111.

Step 7. Our next gross dividend is now 1118. Subtract therefrom $3ac^2+3b^2c=3x6x7^2+3x9x7=882+189$. Now, the actual working dividend is 1118-1071=47. Divide it by 108 and write Q4=0 and R4=47.

Step 8. Our gross dividend is 475. Now subtract $3bc^2=3x7x49=441$. So working dividend is 475-441=34. Therefore, Q5=0 and R5=34.

Step 9. Our last gross dividend now is 343. Subtract from this $c^3=7(3)=343$. Thus Q6=0 and R6=0.

Therefore the number 258474853 is a perfect cube with 637 being its cube root.

Let us apply the above method in finding out cube root of incomplete cubes.

Example 3: Find out cube root of 824 up to three decimal places.

824 = 824 0 0 0

243 95 221 266

9378

Step 1. By inspection first digit of the cube root=9.Therefore divisor=$3a^2$=3x9x9=243. The remainder in this case=95.

824
-729
=95.

Step 2. Gross dividend=950. Do not subtract anything from it and divide it by 243. Q1=3 and R1=221. We have found the cube root up to one decimal place i.e. 9.3.

Step 3. Now the gross dividend=2210. Subtract from it $3a^2$=3 x 9 x 9=243. So the working dividend=1967. Divide it by 243 and write Q2=7 and R2=266. Be careful here. One may be tempted to write Q2=8 but then gross dividend in the next step will be negative. This gives us a clue and we can approximate the cube root as =9·378.

Summary: Procedure for finding cube roots of numbers (exact cube or otherwise) is as follows:

1. Group the digits in the number starting from right.
2. The first digit, number of digits (in case of exact cubes), the division=$3a^2$ and the remainder is available on inspection.
3. From second dividend no subtraction is to be made. Write Q1 and R1 at proper places.
4. From the third gross dividend subtract $3ab^2$ and write Q2 and R2 at appropriate places.
5. From the fourth deduct $6abc+b^3$ and write Q3 and R3.

6. From the fifth subtract $3ac^2=3b^2c$ and write Q4 and R4.

Exercise 2

a) Find cube roots of following numbers. (Exact cubes)

i) 2744 ii) 12167 iii) 39304

iv) 157464 v) 704969 vi) 830584

b) Find cube roots of the following numbers up to two decimal places.

i) 2789 ii) 13250 iii) 39526

iv) 158263 v) 715268 vi) 30628

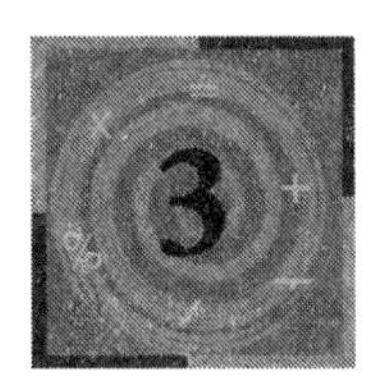

Fractions and Recurring Decimals

In this chapter we shall deal with the conversion of fractions into their equivalent decimal form and demonstrate the superlative vedic methods in dealing with concrete mathematical problems.

After introducing zero notations, the vedic system has been advocating decimal system not because of partiality but solely on its intrinsic merits.

Preliminary Considerations

We begin this part with a brief reference to the well-known distinction between non-recurring, recurring and partly recurring decimals.

1. Non Recurring Decimal Fractions:

A denominator containing only 2 or 5 as factors gives us an ordinary non recurring decimal fraction. Each 2, 5 or 10 contribute one significant digit to the decimal. For instance:

$\frac{1}{2}=\cdot 5$ $\quad \frac{1}{4}=1/2\times 2=\cdot 25$ $\quad 1/8=1/2\times 2\times 2=\cdot 125$ $\quad 1/20=\cdot 05$

$1/5=\cdot 2$ $\quad 1/10=\cdot 1$ $\quad 1/25=1/5^2=\cdot 04$

$1/32=1/2^5=\cdot 03125$ $\quad 1/80=1/10\times 2^3=\cdot 0125$

$1/100=1/10^2=\cdot 01$ and so on.

2. Recurring Decimals:

Denominators containing only 3, 7, 11 or higher prime numbers as factors and not even a single 2 or 5 shall give recurring or circulating decimals which shall be dealt with in great detail in this chapter.

3. Partly Recurring Decimals:

A denominator with factors like 2 and 5 and also factors like 3, 7, 9 shall give us a partly recurring decimal each 2, 5 or 10 contributing one non-recurring digit to the decimal. For example:

$1/6=1/2\text{x}3=\cdot 1\dot{6}$ $1/15=1/3\text{x}5=\cdot 0\dot{6}$ $1/18=1/2\text{x}9=\cdot 0\dot{5}$

$1/22=1/2\text{x}11=\cdot 04\dot{5}$

4. In every non-recurring decimal with the standard numerator i.e. 1, it will be observed that the last digit of the denominator and the last of the equivalent decimal, multiplied together will always yield a product in zero.
5. In every recurring decimal with the standard numerator i.e. 1, it will be similarly observed that 9 will invariably be the last digit of the product of the last digit of the denominator and the last digit of its recurring decimal is actually a continuous series of 9. Thus:

½=·5   1/5=·2   1/10=·1   ¼=·25   1/8=·125

1/25=·04 1/16=·0625

And

$1/3=\cdot\dot{3}$ $1/7=\cdot\dot{1}4285\dot{7}$ $1/9=\cdot\dot{1}$ $1/11=\cdot\dot{0}\dot{9}$

The above property enables us to determine beforehand the last digit of the recurring decimal equivalent of a given fraction. Thus,

i) 1/17 will have a recurring decimal with last digit=7 (the product 7x7 giving 9).
ii) 1/19 will have a recurring decimal with last digit=1 (product of 9x1=9).
iii) 1/21 will have a recurring decimal with last digit=9 (product of 1x9=9).

iv) 1/23 will have a recurring decimal with last digit=3 (product of 3x3=9).

Fractions having their last digit as 9

In this we shall consider fractions of the type 1/19, 1/29, 1/39 and so on. Let us find decimal equivalent of 1/19.

1. Let us consider normal method of division.

19) $\dot{1}$·00(·0526315789473682$\dot{1}$
95
50
38
120
114
60
57
30
19
110
95
150
133
170
152
180
171
90
76
140
133
70
57
130
114

160
<u>152</u>
  80
  <u>76</u>
    40
    <u>38</u>
      20
      <u>19</u>
        1

In the above method 18 long steps are required to convert 1/19 into its equivalent recurring decimal.

2. Now consider the superior one line vedic method using Ekadhika Purva Sutra.

$1/19 = \dot{0}52631578/94736842\dot{1}$

Steps required for the above results are:

Step 1: The fraction 1/19 will have a recurring decimal equivalent.

Step 2: We also know that the last digit of the equivalent recurring decimal has to be 1 (refer to rule 5). So that the product of the last digit of the denominator and the last digit of the recurring decimal = 1x9=9. Step 2 is important as it is the stepping stone for applying Ekadhika method.

Step 3: Now the last but one digit of the denominator is 1; we increase it by 1 and make it 2.

Step 4: Putting 1 as the last digit and continuously multiplying by 2 towards the left we get the last four digits towards the left without the least difficulty.

947368421
11 1

Step 5: 8x2=16. Therefore put 6 down immediately

to the left of 8 with 1 to carry over. Next 6x2=12+1 (carried one)=13. Put 3 to the left with 1 carry over. Next 3x2=6+1 (carry over). Next 7x2=14 put 4 to the left of 7 with 1 carry over. Finally 4x2=8+1(carry over). Put 9 to the left of 4.

Step 6: We have thus got 9 digits by continual multiplication from the right towards the left by 2. Now 9x2=18.Which is same as (D-N) denominator 19-numerator 1. This means half of the work is over and the next 9 digits are obtainable by putting down the complements from 9 of the digits already found and that explains why the slash was put after 9.

Thus the answer is:

1/19=·052631578/947368421

Let us now have some more examples of fractions having their last digit 9 and find their decimal equivalent.

Example 1: Find the value of 1/29.

Step 1. We know that the last digit of the recurring decimal equivalent will be 1 so that the product of last digits of fraction and recurring decimal equivalent is 9x1=9.

Step 2. Using Ekadhika Purva Sutra we multiply the extreme right digit 1 by 3 (2+1=3). So the second right hand digit=3.

Step 3. Now we go on multiplying by 3 and carry over the surplus digit or digits if any, to the left. Thus we have obtained 14 digits i.e.

96551724137931

1 1 2 1 1 2

Step 4. The next product is 9x3+1=28. Since 28=(D-N)=(29-1)=28. We know that we have completed half of the work.

Step 5. Now we set the first 14 digits by simply subtracting each of the above digits from 9.

Thus 1/29=·03448275862068/96551724137931

Example 2: Next let us consider 1/39.

Step 1. Take again 1 as the extreme right digit of the recurring decimal.

Step 2. Applying Ekadhika Purva Sutra we keep multiplying with 4 (3+1). Thus we have

1/39=·$\dot{0}2564\dot{1}$

        2 1

Step 3. Note here 39=3x13 and is not a prime number like 19 or 29.

The factors 3 and 13 give only 1 (1/3=·0$\dot{3}$) and 6

(1/13=·$\dot{0}7692\dot{3}$) recurring decimals.

Step 4. Therefore 1/39 has only 6 digits in its recurring decimal equivalent and thus

1/39=·$\dot{0}2564\dot{1}$

Example 3: Now consider 1/49.

Step 1. The extreme right hand digits of the recurring decimal in this case is=1.

Step 2. Applying Ekadhika method we keep multiplying from right to left by 5. After completing 21 steps, i.e.

979591836734693877551

3244133123414332 2

Step 3. On completing 21 digits we find 48 i.e. (D-N=49-1=48) coming up staring at us and we mechanically put down the other 21 digits as usual by subtraction from 9.

The answer is

$1/49 = .\dot{0}20408163265306122448/97959183673469387755\dot{1}$

Fractions having last digit of the denominator other than 9.

In this we now go on to study the cases of 1/7, 1/13, 1/17, 1/23 and other such fractions whose denominator ends in digits other than 9 but with digits 1, 3 or 7. In such cases first we make up our minds by inspection regarding the last digit of the decimal equivalent. Thus, denominator ending with 7, 3 and 1 necessarily have decimal equivalent last digits as 7, 3 and 9 so that the product of the last digit of the denominator and the last digit of decimal equivalent is equal to 9.

Example 4: Consider 1/7.
First method: Using Ekadhika Purva method for multiplication.

Step 1. Put down 1/7=7/49.

Step 2. Take 5 i.e. 1 more than 4 for the required multiplication.

Step 3. Thus starting with 7 at the right end we get the three digits as shown: 8 5 7
2 3

Step 4. On fourth multiplication we get 8x5+2=42 which is same as (D-N) i.e. (49-7=42).

Step 5. Stop at the fourth multiplication and get the first three digits from the complement rule as 142.
The answer is

$1/7 = .\dot{1}4285\dot{7}$

Second method: Using Ekadhika Purva method for division.

Step 1. The Ekadhika being 5, divide 7 by 5 and continue the division as usual. Q=1 and R=2 are written as shown

1

2

Step 2. 21 on division by 5 gives Q=4 and R=1 and is written as:

1 4 2

2 1 4

Step 3. 14 on division by 5 gives Q=2 and R=4 and they are written as shown above.

Step 4. Now the dividend=42, which is same as (D-N) i.e. (49-7=42). We stop further division and write the other three digits using the complement rule and get the answer

$$1/7=7/49=\dot{1}4285\dot{7}$$

Example 5: Consider 1/13

Step 1. Firstly 1/13=3/39.
The right hand most digit of the decimal equivalent shall be 3.

Step 2. Ekadhika digit for multiplication or division = 3+1=4.

First method: Using Ekadhika Purva method for multiplication.

Step 3. Multiplying with 4 we get the three digits as shown

9 2 3

1

Step 4. Further multiplication yields 9x4=36 which is same as (D-N)
=39-3=36.

Step 5. Stop further multiplication and other three digits can be mechanically put down as complement of 9 as 076.

Answer can be written as

. .

1/3=3/39=·076923

Second method: Using Ekadhika Purva method for division.

Step 1. Divide 3 by 4. Q=0 and R=3.

Step 2. Divide 30 by 4. Q=7 and R=2.

Step 3. Next dividend 27 divided by 4 gives Q=6 and R=3.

.0 7 6
3 2 3

Step 4. Since we got dividend=36 we stop now and get other three digits as complement of 9. Answer can be written as

. .

1/13=3/39=076923

Example 6: Consider 1/11.

Step 1. 1/11=9/99

Step 2. The last digit of the equivalent decimal is 9.

Step 3. Ekadhika=10

Step 4. After the right hand most digit is 9 on multiplication with 10 yields 9x10=90 which is the same as (D-N) i.e. 99-9=90.

Step 5. We stop further multiplication and write the answer as:

. .

1/11=9/99=·09

Example 7: Consider 1/23.

Step 1. 1/23=3/69

The right hand most digit is 3 and Ekadhika=7.

Step 2. On continuos multiplication we get:

695652173913
63431152612

Step 3. After 11 steps we get 66 which is (D-N) i.e. 69-3=66. We stop here and using complement of 9 write the answer:

1/23=3/69=·04347826086/95652173913

Example 8: Consider 1/17.

Step 1. 1/17=7/119

The last digit of the equivalent decimal is 7 and Ekadhika=12.

Step 2. On multiplication we get the eight right hand most digits as

294117647

114129758

Step 3. We stop further multiplication as we get 112 which is same as (D-N)=112.

Step 4. Getting the other digits as complements of 9 we get the answer:

$1/17=7/119=\dot{0}5882352/9411764\dot{7}$

Multiples of the basic fractions

So far we have dealt with fractions whose numerator is unity, but what about fractions which have some other numerators? The answer is that there are several simple and easy methods by which we can get the result. Let us consider 1/7.

$1/7=\cdot\dot{1}4285\dot{7}$

$2/7=\cdot\dot{2}8571\dot{4}$

$3/7=\cdot\dot{4}2857\dot{1}$

$4/7=\cdot\dot{5}7142\dot{8}$

$5/7=\cdot\dot{7}1428\dot{5}$

6/7=·857142

In the above we observe that

a) The same six digits are there as in the case of 1/7.
b) They come up in the same sequence and in the same direction as in case of 1/7.
c) They, however start from a different starting point but in 'cyclic' order in clockwise direction.
d) With the aid of the above rules it is easy to find the equivalent recurring decimal of a fraction whose numerator is greater than 1.

An independent method:

It is not mandatory to remember the above tabulated results in order to find the decimal equivalent of a fraction whose numerator is greater than 1. Consider 3/7=21/49. In this case Ekadhika will be 5. Dividing 21 by 5 we get:

·4 2 8
1 4 2

And since we get 28 as the dividend which is (D-N)=(49-21)=28, we stop and write the other three digits, 571, as complement of 9 and thus the answer is:

3/7=21/49=·428/571
1 4 2——— Remainder at each stage.

Exercise 3

Find decimal equivalent of the following fraction.

i) 5/7	ii) 3/13	iii) 4/19	iv) 4/7
v) 1/11	vi) 6/11	vii) 4/13	viii) 3/49
ix) 5/59	x) 3/29	xi) 5/39	xii) 2/7

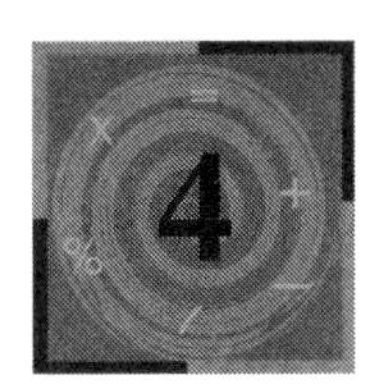

Auxiliary Fractions: a Method For Fast Calculation

Conventional Method:

While finding decimal value of a fraction we manipulate the decimal point. Thus:

i) 1/700=·01/7 ii) 39/70=3·9/7 iii) 17/130=1·7/13

iv) 3741/11000=·3741/11 v) 97654/90000000=·0097654/9 etc.

But after this has been done the other operation of actual division is to be carried in the usual manner.

Vedic Method for Auxiliary Fraction:

There are certain vedic processes by which the burden of actual division is considerably lightened and the work is easily done. These methods are given below:

First method: Denominators ending with 9

Example 1: Find 73/139 up to 6 places of decimal.

The conventional method will involve:

```
139) 730 (·525179
     695
      350
      278
       720
       695
        250
        139
```

1110
973
1370
1251
119

Now we will solve the above example with the help of Auxiliary Fraction.

Step 1. 73/139 is reduced to:

73/139=7·3/13·9=7·3/14

Step 2. Start dividing 7·3 by 14.

Step 3. Put the decimal point and divide 73 by 14. Q=5 and R=3. 5 is written after the decimal and 3 is written in front of 5 as shown . The Remainder is brought forward:

0· 5
3

Step 4. Next gross dividend is 35. Divide 35 by 14 and write Q=2 and R=7 are written as shown below:

Step 5. Next gross dividend 72 on division with 14 gives Q=5 and R=2 are written as below:

0. 5 2 5
3 7 2

Step 6. Continue with the process till we get decimal values up to 6 places. Answer is:

73/139=0. 5 2 5 1 7 9
3 7 2 11 13 11

We can see how the whole process has been simplified.

Example 2: Find 63/149 up to 4 decimal places.

Step 1. Denominator ending with 9

63/149=6·3/14·9=63/15

Step 2. Divide 6·3/15. Place decimal and Q=4 and R=3 as

. 4
3

Step 3. Continue with the process up to four decimal places and write the answer as:

```
. 4 2  28
 3 4  12
```

Example 3: Find the value of 1/19 up to 5 decimal places. This problem has already been dealt with in the previous chapter of recurring decimal.

The value of $1/19 = \cdot\dot{0}52631578/9473684\dot{1}$

Now we shall find its value up to five decimal places using auxiliary fractions.

Step 1. 1/19=1/2. Divide 1 with 2. Put Q=1 and R=1.

Step 2. Place decimal and divide 10 with 2. Q=5 and R=0.

Step 3. Dividend is 05 which, on division with 2 gives Q=2 and R=1 and they are placed as shown.

Step 4. Gross dividend is 12 and therefore Q=6, R=0.

```
. 0 5 2 6 3
 1 0 1 0 0
```

Step 5. Gross dividend is 06 and therefore Q=3, R=0 So the answer is

1/19=·05263

Example 4: Find 75/139 up to five decimal places.

Step 1. 75/139=7·5/13·9=7·5/14

Step 2. Place decimal and divide 75 with 14. Q=5 and R=5

```
. 5  3 9 5  6
 5 13 7 9  11
```

Step 3. Next dividends are 55, 133, 79, 95 and 116 which on division with 14 give Q=5 and R=5, Q=3 and R=13, Q=9 and R=7, Q=5 and R =9 and Q=6 and R=11 respectively.

The answer is:

75/139=·53956

Compare this with the troublesome process of division.

```
139) 750 (·53956
     695
      550
      417
      1330
      1251
        790
        695
         950
         834
         116
```

You will find the answers in both the cases are the same.

Example 5: Find 63/149 up to five decimal places.

Step 1. 63/149=6·3/14·9=6·3/15

Step 2. Place decimal and divide 63 with 15. Write Q=4 and R=3.

Step 3. Next dividends are 34, 42, 122, 28 and 131 respectively which on division with 15 will give Q=2 and R=4; Q=2 and R=12; Q=8 and R=2; Q=1 and R=13.

```
. 4  2   2  8   1
 3  4  12  2  13
```

The answer is

63/149=·42281

Let us compare with actual division and see whether the answer is same.

```
149) 630 (·42281
     596
      340
      298
       420
       298
```

1220
<u>1192</u>
 280
 <u>149</u>
 131

We see that the answers are exactly the same and the Vedic method of auxiliary fractions is much superior.

Example 6: Find 83/189 up to five decimal places.

Step 1. Denominator ending at nine.
83/189=8·3/18·9=8·3/19

Step 2. Start dividing 83 by 19, Q=4 and R=7.

Step 3. Dividend now is 74 divided by 19 gives Q=3 and R=17. Write them as shown below:

. 4 3 9 1 5
7 17 2 10 6

Step 4. Subsequent dividends are 173, 29, 101 and 65 which on division with 19 yield Q=9 and R=2, Q=1 and R=10, Q=5 and R=6 respectively.
The answer is
83/189=·43915

Denominators ending with 8, 7 and 6

The above method where the denominator ended at 9 can also be applied with light modification to fractions where the last digit is not 9 but 8, 7 and 6. Let us study these cases with some examples.

Example 1: Find 89/138 up to five decimal places.

Step 1. 89/138=8·9/13·8=8·9/14.
Place decimal and divide 89 with 14. Write Q=6 and R=5. Put R in front of Q.
$8.9/14=5^{6+6}6^{4+4}12^{4+4}2^{9+9}10^{2+2}$

Step 2. Now the gross dividend=56+6(9-last digit)
=56+6(9-8)
=56+6
=62

Step 3. 62 on division with 14 gives Q=4 and R=6.

Step 4. Gross dividend=64+4=68 which gives Q=4 and R=12.

Step 5. Gross dividend=124+4=128 Which gives Q=9 and R=2.

Step 6. Gross dividend=29+9=38 which gives Q=2 and R=10.

Answer therefore is

89/138=·64492

Compare and contrast this with the conventional division method.

138) 890 (·64492
828
620
552
680
552
1280
1242
380
276
104

You must notice here that though we converted 89/138 to an auxiliary fraction 8·9/14 the answer in both the cases is exactly the same. What is being emphasized is that the auxiliary fraction method is not an approximate method.

Example 2: Consider another example of a fraction terminating with 8 as the last digit. Find 76/118 up to five decimal places.

Step 1. 76/118=7·6/11·8=7·6/12. Since the denominator ends with 8. The quotient digit x (9-8=1) is to be added on to the quotient digit.

Step 2. Place decimal and divide 76 by 12. Q=6 and R=4

7·6/12=. 6+6 4+4 4+4 0+0 6+6
4 4 0 8 8

Step 3. Now gross dividend=46+6=52 and not 46. On division with 12 gives Q=4 and R=4.

Step 4. Gross dividend 44+4=48 on division gives Q=4 and R=0.

Step 5. Gross dividend 04+4=8. Q=0 and R=8.

Step 6. Gross dividend 80+0=80. Q=6 and R=8. Answer is 76/118=·64406

The actual division is:

```
118) 760 (·64406
     708
      520
      472
       480
       472
         800
         708
          92
```

Answers in both the cases are the same.

Example 3: Find 33/188

Step 1. Denominator ending with 8 therefore the Quotient digit is to be added to it. 33/188=3·3/18·8=3·3/19

Step 2. Put decimal and divide 33 with 19. Q=1 and R=14.

3·3/19=. 1+1 7+7 5+5 5+5 3+3
14 9 9 5 3

Step 3. Next gross dividend is 141+1=142 which on division with 19 gives Q=7 and R=9.

Step 4. Gross dividend is 97+7=104 which on division gives Q=5 and R=9. Next we divide 95+5=100 and get Q=5 and R=5. Now dividend=55+5=60 which yields Q=3 and R=3.

We can write the answer:

33/188=·177553 which is exactly the same as obtained by division given below.

```
188) 330 (·177553
     188
     1420
     1316
       1040
        940
        1000
         940
           600
           564
            36
```

Denominator having last digit as 7

Now let us consider denominator ending with 7. In these cases the quotient digit is increased by (9-7)=2 i.e. twice. Procedure will be clear from the following examples.

Example 1: Find 17/127 up to five decimal places.

Step 1. 17/127=1·7/12·7=1·7/13, denominator ending with 7.

Step 2. Put decimal and divide 17 with 13. Q=1 and R=4. Add 1x2=2 i.e. twice the quotient digit to get new dividend as shown below:

```
17/127=1·7/13=. 1+2  3+6  3+6  8+6  5
                  4    4   10    5   9
```

Step 3. New dividend is 41+2=43 on division with 13 gives Q=3 and R=4.

Step 4. Next dividend is 43+6=49 which on division gives Q=3 and R=10. Now dividend is 103+6=109, on division we get Q=8 and R=5. Dividend now is 58+16=74 and will give Q=5 and R=9.

The answer is:

17/127=·13385

Let us compare this with actual division.

```
127) 170 (·13385
     127
      430
      381
       490
       381
       1090
       1016
        740
        635
        105
```

Here again we see that the results are exactly the same. This highlights the ease of auxiliary fraction.

Example 2: Find 93/177 up to five decimal places.

Step 1. 93 /177=9·3/17·7=9·3/18 denominator ending 7.

Step 2. Put decimal and divide 93 by 18. Q=5 and R=3.

```
93/177=9·3/18=. 5+10  2+4  5+10  4+8  2+4
                  3     9     6     3    6
```

Step 3. Continuing as before denominator 35+10=45. On division with 18 gives Q=2 and R=9. Next denominator 92+4=96, which on division gives Q=5 and R=6. Next denominator 34+8=42, Q=2 and R=6.

We can write the answer as:

93/177=·52542

Let us check the answer with actual division.

```
177) 930 (·52542
     885
      450
      354
       960
       885
        750
        708
         420
         354
          66
```

Here again we see that the answers are exactly the same.

Example 3: Find 29/157 up to five decimal places.

Step 1. 29/157=2·9/15·7=2·9/16, denominator ending in 7.

Step 2. Place decimal and divide 29 with 16
Q=1 and R=13

```
29/157=.  1+2   8+16   4+8   7+14   1+2
           13     5     10     0      5
```

Step 3. Dividend is 131+2=133. On division with 16 it gives Q=8 and R=5

Step 4. Next dividend is 58+16=74. On division with 16 it gives Q=4 and R=10. Now dividend is 104+8=112, which on division with 16 gives Q=7 and R=0

Step 5. Now the next dividend is 7+14=21. Division with 16 yields Q=1 and R=5. We stop here and write the answer.

29/157=2·9/15·7=2·9/16=·18471

Let us check this with actual division:

```
157) 290 (·18471
     157
     1330
     1256
        740
        628
        1120
        1099
         210
         157
          53
```

Here again the examples are exactly the same.

Denominators having last digit as six:

Now we consider denominator ending with 6. In such cases the quotient digit is increased by (9-6)=3 i.e., thrice its value. This is well illustrated with few examples.

Example 1: Find 73/126 up to five decimal places.

Step 1. 73/126=7·3/12·6=7·3/13, last digit of denomination ending in 6.

Step 2. Place decimal and divide 73 by 13 gives Q=1 and R=13

```
73/126=7·3/13=. 5+15 7+21 9+27 3+9 6+18
                  8     9     1    7    4
```

Step 3. Next dividend is 85 + 3 x 5=100. On division with 17 gives Q=7 and R=9. New dividend is 97+7 x 3=118. On division yields Q=9 and R=1.

Step 4. Now dividend is 19+27=46, which on division with 13 gives Q=3 and R=7. We get dividend 73+9=82, which on division with 16 gives Q=6 and R=4. We stop here and write the answer.

Step 5. The answer can be written as
73/126=·57936

Let us check this with actual division:

126) 730 (·57936
630
1000
882
1180
1134
460
378
820
756
64

And we find that answers are exactly the same.

Example 2: Find 83/156 up to five decimal places.

Step 1. 83/156=8·3/15·6=8·3/16, last digit of denomination ending in 6.

Step 2. Place decimal and divide 83 by 16 which gives Q=5 and R=3. Write them as shown. Next dividend 35+15=50

83/156=8·3/16=.	5+15	3+9	2+6	0+0	5
	3	2	0	8	0

Step 3. Divide 50 with 16, which gives Q=3 and R=2. New dividend is 23+9=32. On division this gives Q=2 and R=0.

Step 4. Now dividend is 8, which on division with 13 gives Q=0 and R=8. Next dividend is 80, which on division with 16 gives Q=5 and R=0. We stop here and write the answer.

Step 5. The answer is
83/156=·53205

Let us check this with actual division:

156) 830 (·53205
780
500
468
320
312
800
780
20

We find that the answers are exactly the same.

Example 2: Find 96/186 up to five decimal places.

Step 1. 96/186=9·6/18·6=9·6/19, last digit of denomination ending in 6.

Step 2. Place decimal and divide 96 by 19, which gives Q=5 and R=1. Write them as shown. Next dividend

96/186= . 5+15 1+3 6+8 1+3 2+6
1 11 0 5 16

Next dividend 15+15=30, on division with 19 will give Q=1 and R=11. New dividend is 111+3=114. On division gives Q=6, R=0.

Step 3. Now dividend is 24. This will give Q=1 and R=5. Next dividend is 51+3=54. On division this gives Q=2 and R=16.

Step 4. The answer is

96/186=·51612

Let us check this with actual division:

186) 960 (·51612
930
300
186
1140
1116

240
186
540
372
168

Here again answers are exactly the same.

Auxiliary fraction using Eknyuna Method

So far we have learned to handle denominates ending with 9, 8, 7 and 6. These follow Ekadhikena Purvena Sutra.

For denominators ending in 1 we follow Eknyuna method. It is same as Ekadhikena Purvena, where our gross dividend used to be Remainder Quotient but in this case gross dividend is Remainder (9-Quotient). This method will be clear with following examples.

Example 1: Find 72/121 up to five decimal places.

Step 1. Subtract 1 from denominator as well as numerator 72/121=72-1/121-1=71/120=7·1/12.

Step 2. Place decimal and divide 71 by 12, gives Q=5 and R=11. Write them as shown below

72/121=7·1/12= 54 9 54 09 45 18
11 6 0 4 1 3

Step 3. Subtract quotient digit from 9 and write 9-5=4 as shown. Gross dividend now is 114, which on division with 12 gives Q=9 and R=6. Write them as before.

Step 4. Now gross dividend=60 and it will give Q=5 and R=0. Gross dividend is 4. On division with 12 gives Q=0 and R=4.

Step 5. Gross dividend is 49. On division with 12 will give Q=4 and R=1. Next gross dividend is 15 which gives Q=1 and R=3. Stop at this and write the answer.

72/121=·59504

Let us check this with actual division:

121) 720 (·59504
605
1150
1089
610
605
500
484
16

Here again we find that both the answers are the same.

Example 2: Find 57/171 up to five decimal places.

Step 1. Subtract 1 from the denominator as well as the numerator 57/171=57-1/171-1=56/170=5·6/17.

Step 2. Place decimal and divide 56 by 17, which gives
Q=3 and R=5. Write them as shown below
57/171=5·6/17=· 36 36 36 36 36
5 5 5 5 5

Step 3. Write (9-3)=6 near the Q digit. Dividend now is 56, which on division with 17 gives Q=3 and R=5. This is a case of recurring decimal and we can write the answer.
57/171=·33333

Let us check with actual division:

171) 570 (·3333
513
570
513
570
513

570
513
570
513
570
....

And we can see that the answers are exactly the same.

Example 3: Find 73/141 up to five decimal places.

Step 1. Subtract 1 from the denominator as well as the numerator 73/141=73-1/141-1=72/140=7·2/14.

Step 2. Place decimal and divide 72 by 14, gives Q=5 and R=2.

7·2/14= ·54 18 72 72 36
2 10 10 4 0

Step 3. Carry on as before and write the answer 73/141=·51773.

Let us check this with actual division:

141) 730 (·51773
705
250
141
1090
987
1030
987
430
423
70

Again the answers are exactly the same.

Numerators having more than one digit after decimal

So far, we have dealt with problems wherein the numerator, the number of digits after the decimal was 1 like in 72/121=72-1/121-1=71/120=7·1/12. Number of digit after decimal is 1.

Does the same rule apply when the number of digits after the decimal is more than 1?

The answer is Yes. We can apply the same rules and technique as before. But remember, the remainder will be brought forward after completion of exactly the same number of operations as the number of digits after the decimal. Consider following examples.

Example 1: Find 738/1399 up to five decimal places.

Step 1. 738/1399=7·38/13·99=7·38/14. Denominator ending in 9

7·38/14= 52 75 19
10 2 9

Step 2. Place decimal and divide 73 by 14, gives Q=5 and R=3. As explained earlier do not bring forward the remainder now, instead divide new dividend 38 by 14, which gives Q=2 and R=10.

Step 3. Following the two digit cycle dividend now is 105, which on division with 14 gives Q=7 and R=7. Divide 72 with 14, Q=5 and R=2.

Step 4. Bring forward the remainder 2. Now dividend=27 and it will give Q=1 and R=13. Next dividend is 135, on division with 14 gives Q=9 and R=9. Bring forward this remainder after completion of two cycles of divisions, which is the same as the number of digits after the decimal.

Step 5. The answer is
738/1399=·527519

Let us check this with actual division:

```
399) 7380 (·527519
     6995
      3850
      2798
      10520
       9793
        7270
        6995
         2750
         1399
         13510
         12601
           909
```

We find that the answers are the same.

Example 2: Find 326/889 up to five decimal places.

Step 1. 326/899=3·26/8·99=3·26/9. Denominator ending in 9

Step 2. Place decimal and divide 32 by 9, gives Q=3 and R=5. Since the numerator has two places of decimal we shall undertake cycle of division twice before bringing the remainder forward. Therefore next dividend is 56, which gives Q=6 and R=2. Thus, we write

```
326/899=3·26/9=· 36 26 25
                  2  2  1
```

Step 3. The answer can be written
326/899=·362625

Let us check this with actual division:

```
899) 3260 (·36262
     2697
      5630
      5394
       2360
       1798
        5620
        5394
         2260
         1798
          462
```

We find that the answers are exactly the same.

Example 3: Find 1400/1401 up to eight decimal places.

Step 1. Using Eknyuna method
1400/1401=1400-1/1401-1=1399/1400=13·99/14 (with a group of two digits).

Step 2. Since the numerator has a two-digit group after the decimal we shall take the division cycle twice before bringing forward the remainder.

Step 3. Place decimal. Divide 139 by 14. Q=9 and R=13. New dividend 139 which on division with 14 gives, Q=9 and R=13. Bring forward the remainder as shown:

```
1400/1401=13·9/14=·  99  92  86  22
                   13  12   3   5
```

Step 4. Next dividend is not 139 but 13(9-9) i.e. complement of the quotient digit=130. Division by 14 gives Q=9 and R=4. Next dividend is 4(9-9)=40 (here again the second digit is not 9 but complement of 9 i.e. 0). On division it gives Q=2 and R=12.

Step 5. Next dividend is 12(9-8)=120. On division with 14 gives Q=8 and R=8. Following the division cycle twice our new dividend is

8(9-2)=87, Q=6 and R=3 which is brought forward. Carrying forward new dividend is 3(9-8)=31 giving Q=2 and R=3. Next dividend is 3(9-6)=33 yielding Q=2 and R=5. The answer is: 1400/1401= Auxiliary fraction=13·99/14

=·99928622

Let us check this with actual division.

```
1401) 14000 (·99928622
      12609
       13910
       12609
        13010
        12609
          4010
          2802
         12080
         11208
           8720
           8406
            3140
            2802
             3380
             2802
              578
```

We find the answers to be exactly the same.

Example 4: Find 131/701 up to six decimal places.

Step 1. Utilizing Eknyuna method 131/701=131-1/701-1=130/700=1·30/7 (with group of two digits).

Step 2. Place decimal and Divide with 7. Q=1 and R=6. Next dividend 60 which on division with 7 gives, Q=8 and R=4. write:

131/701=1·30/7=·18 68 75
4 5 6

Step 3. Next dividend is 4(9-1), the second digit is complement of 9. On division 48 gives Q=6 and R=6. Now dividend is 6(9-8)=61. On division it gives Q=8 and R=5.

Step 4. Next dividend is 5(9-6)=53. On division with 7 gives Q=7 and R=4. Our next dividend is 4(9-8)=41, Q=5 and R=6.

We can write the answer:

131/701= ·186875

Let us check this with actual division.

```
701) 1310 (·186875
      701
      6090
      5608
       4820
       4206
        6140
        5608
         5320
         4807
          5130
          3505
          1625
```

We find the answers to be exactly the same.

Sometimes the fraction apparently not in the Eknyuna format, can be converted into one. This is illustrated in the following example.

Example 5: Find 5/67 up to six decimal places.

Step 1. 5x3/67x3=15/201=15-1/201-1=14/200= ·14/2 (with a group of two digits).

Step 2. Place decimal and divide with 2. Q=0 and

R=1. Next dividend is 17 which on division with 2 gives, Q=7 and R=0. Write:

5/67=15/201=·14/2 =· 07 46 26
0 0 1

Step 3. Next dividend is 0(9-0). This gives Q=4 and R=1. Next dividend is 1(9-7)=12 giving Q=6 and R=0.

Step 4. Next dividend is 0(9-4)=5, gives Q=2 and R=1. New dividend is 1(9-6)=13, Q=6 and R=1.

The answer is:

5/67= ·074626

Let us check this with actual division.

```
67) 500 (·074626
    469
     310
     268
      420
      402
       180
       134
        460
        402
         58
```

Both the answers are exactly the same.

Example 6: Find 31/77 up to five decimal places.

Step 1. 31x3/77x3=93/231=93-1/231-1=92/230=9·2/23 (with group of two digits).

Step 2. Put decimal and divide 92 by 23. Q=4 and R=0. Next dividend is 05 giving Q=0 and R=5. write:

31/77=9·2/23 =· 45 09 27 54 9
0 5 13 22 17

Step 3. Divide 59 with 23. Q=2 and R=13. Next dividend 137 on division with 23 yields Q=5 and R=22. Divide 224 with 23, Q=9 and R=17.

Step 4. The answer can be written as.

31/77= ·40259

Let us check this with actual division.

```
77) 310 (·40259
    308
    ---
      200
      154
      ---
       460
       385
       ---
        750
        693
        ---
         57
```

Here again both the answers are exactly the same.

Now consider another example where we have to make three-digit cycle of division before moving the remainder forward.

Example 7: Find 2743/7001 up to nine decimal places.

Step 1. 2743/7001=2743-1/7001-1=2742/ 7000=2·742/7 (with three digit group).

Step 2. Carry out a cycle of three divisions before bringing forward the remainder. Write down complements of quotient digit at the top for ease:

```
                          6 0 8   1 9 8   7 2 8
2743/7001=2·742/7 = ·  3 9 1   8 0 1   2 7 1
                      5       1       1
```

Step 3. The answer can be written as.

2743/7001= ·391801271

Let us check this with actual division.

7001) 27430 (·40259
<u>21003</u>
64270
<u>63009</u>
12610
<u>7001</u>
56090
<u>56008</u>
8200
<u>7001</u>
1990
<u>1402</u>
5880
<u>4907</u>
9730

Here again we find that the answers are the same.

Example 8: Find 29/15001 up to nine decimal places.

Step 1. 29/15001=29-1/15001-1=28/15000=·028/15. As there are three digits after the decimal we will carry 3-division cycle before bringing forward the remainder.

Step 2. After three-division cycle, we can write the answer:

998 066 795
29/15001=·028/15=· 001 933 204
13 3 6

Step 3. The answer can be written as
29/15001= ·001933204

Let us check this with actual division.

15001) 29000 (·001933204
<u>15001</u>
139990
<u>135009</u>

49810
<u>45003</u>
48070
<u>45003</u>
30670
<u>30002</u>
66800
<u>60004</u>
6896

The answers are the same.

Example 9: Find 137/13000001 up to nine decimal places.

Step 1. 137/13000001=137-1/13000001-1 =136/ 13000000=·000136/13.
Here we will carry six-division cycle before bringing forward the remainder.

Step 2. After three-division cycle, we can write the answer:

999989 4 61539
137/13000001=·000136/13=· 000010 538460
6 9

Step 3. The answer can be written as.
137/13000001= ·000010538460

Let us check this with actual division.

13000001) 13700000 (·000010538460
<u>13000001</u>
69999900
<u>65000005</u>
49998950
<u>39000003</u>
109989470
<u>104000008</u>
59894620
<u>52000004</u>

78946160
78000006
94615400

So the answers in both the cases are the same.

Exercise 4

1. Find the value of the following fractions up to five decimal places:

i) 1/39	ii) 76/139	iii) 42/349
iv) 56/379	v) 28/469	vi) 67/689

2. Solve the following fractions with denominator ending in 8 up to three decimal places.

i) 38/278	ii) 49/388	iii) 67/498
iv) 47/668	v) 98/788	vi) 23/488

3. Find values of the following fraction up to six decimal values:

i) 73/377	ii) 43/477	iii) 69/597
iv) 47/466	v) 35/676	vi) 83/896
vii) 32/131	viii) 95/951	ix) 83/671

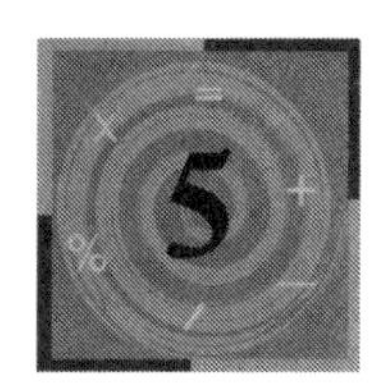

Result Verification Technique: Navashesh Method

As mathematical computations are exact in nature, it is very important to verify the answer obtained. Such techniques are known as result verification techniques. There are various methods, Vedic mathematics offers a very simple method which can be used with great ease.

Navashesh Method: Navashesh means eliminating nine and retaining other digits. Every number has got its unique single digit value. This single digit value is called Navashesh. The Navashesh of any number is the sum of all digits, continued until there is one digit left. Here are some examples.

Number	Summing digits	Navashesh
71	7+1=8	8
231	2+3+1=6	6
85	8+5=13, 1+3=4	4
7562	7+5+6+2=20, 2+0=2	2

It is also a fact that the Navashesh of a number is the same as the remainder when that number is divided by 9. For example 71/9=7, remainder 8, which is Navashesh of 71. Similarly 231/9=25 and remainder 6, which is Navashesh of 231.

Easy way to find Navashesh: An easy method of finding the Navashesh is to cast out nines and group of digits, which add up to 9. This is done by crossing out nines in the number or any digits adding up to nine. The numbers, which are left at the end, are added up for Navashesh. The sutra Navashesh used here means by elimination and retention.

Example 1: Find Navashesh of 19462785.
Step 1. Cross out 9.
Step 2. Cross out (1+8=9); (4+5=9); (2+7=9).
Step 3. The only number left is 6 and therefore Navashesh of 19462785=6

If there is nothing left then the Navashesh=9

Exercise 5

1. Find the Navashesh of the following numbers.

i) 6322	ii) 897364	iii) 367425	iv) 813
v) 43701	vi) 230098	vii) 994652	viii) 657483
ix) 98076	x) 812763		

Navashesh of a negative number:

It must be emphasized the Navashesh of a number whether –ve or +ve is always positive. For instance Navashesh (NV for short) of –732 is –3. But since it can not be negative we add up 9 to make it +ve.

Thus NV of –732=-(7+2+3)=-3=-3+9=6

The Navashesh method can be utilized in verifying the results. This formula simply states that the Navashesh remains unchanged. In other words Navashesh of the digits before operation and after operation will remain unchanged.

Navashesh method for verifying addition operations.

Example 1: 67+34=101
Steps 1. Navashesh of left hand side.
67 = 6+7=13=1+3=4
34 = 3+4=7
7+4=11 = 1+1=2

Step 2. Navashesh of right hand side.
101=1+0+1=2

Step 3. Navashesh of left hand side = Navashesh of right hand side. The answer is correct.

Example 2: 3673+2341=6014

Step 1. Navashesh of left hand side.
Cross out 9. 3673=7+3=10=1
+2341=1=1
Therefore 1+1=2

Step 2. Navashesh of right hand side.
6014=6+0+1+4=11=2

Step 3. Navashesh of left hand side = Navashesh of right hand side. The answer is correct.

Example 3: 3251+6242+845=10338

Step 1. Navashesh of left hand side.
Cross out 9. 3251=2=2
+6242=14=5
+ 845=8=8
Therefore 2+5+8=15=1+5=6

Step 2. Navashesh of right hand side.
Cross out 9.10338=0+3+3=6

Step 3. Navashesh of left hand side = Navashesh of right hand side. The answer is correct.

Example 4: 854+564+3254+12+6524=10938

Step 1. Navashesh of left hand side.
Cross out 9. 854=8
+564=6
+3254=3+2=5
+12=1+2=3
+6524=6+2=8
Therefore 8+6+5+3+8=30=3+0=3

Step 2. Navashesh of right hand side.
Cross out 9.10938=0+3=3

Step 3. Navashesh of left hand side = Navashesh of right hand side.
The correctness of the answer is verified.

Navashesh method for verifying subtraction.
As in the case of addition, Navashesh method can be applied for subtraction with the same rule i.e. Navashesh of left-hand side = Navashesh of right hand side. Let us consider a few examples.

Example 1: 88-31=57

Step 1. Navashesh of left side.
88=8+8=16=1+6=7
-31=3+1=4
Therefore, 7-4=3

Step 2. Navashesh of right side.
57=5+7=12=1+2=3

Step 3. Navashesh of left hand side = Navashesh of right hand side. The answer is correct.

Example 2: Verify the answer of the following subtraction.

5283-2413=2870

Step 1. Navashesh of left side.
5283=5+2+8+3=18=1+8=9
Cross out 9.-2413=-1
Therefore, 9-1=8

Step 2. Navashesh of right side.
Cross out 9.2870=8+0=8

Step 3. Since Navashesh of left hand side = Navashesh of right hand side, the answer is verified.

Example 3: Calculate 7831-4211-1439+152+6524 and check the answer 8857 for correctness.

Step 1. Navashesh of left side.
Cross out 9.

7831=7+3=10=1+0=1
-4211=-(4+2+1+1)=-8
-1439=-(1+4+3)=-8
+152=1+5+2=+8
+6524=6+2=+8
Therefore, 1-8-8+8+8=1

Step 2. Navashesh of right side.
8857=8+8+5+7=28=2+8=10=1+0=1

Step 3. Since Navashesh of left hand side = Navashesh of right hand side the answer is correct.

Example 4: Check the correctness of the following
7831-4211-1439-152-6524=-4495

Step 1. Navashesh of left side.
Cross out 9.
7831=7+3=10=1+0=1
-4211=-(4+2+1+1)=-8
-1439=-(1+4+3)=-8
-152=1+5+2=-8
-6524=6+2=-8
Therefore, 1-8-8-8-8=-31=-(3+1)=-4

Step 2. Navashesh of right side.
Cross out 9.-4495=-(4+4+5)=-13=-(1+3)=-4

Step 3. Navashesh of left hand side = Navashesh of right hand side. Therefore the answer is correct.

Navashesh method for verifying multiplication.

The same rule as in case of addition and subtraction also applies for multiplication i.e. the Navashesh of left hand side = Navashesh of right hand side. Let us consider a few examples.

Example 1: Verify the result of the following multiplication by Navashesh method.

73x51=3723

Step 1. Navashesh of left side.
Cross out 9 if any.
73=7+3=10=1+0=1
51=5+1=6
Therefore, 1x6=6

Step 2. Navashesh of right side.
Cross out 9.3723=3+3=6

Step 3. Navashesh of left hand side = Navashesh of right hand side. Correctness of the answer is verified.

Example 2: Verify the following multiplication
526x812=427112

Step 1. Navashesh of left side.
Cross out 9 if any.
526=5+2+6=13=1+3=4
812=2
Therefore, 4x2=8

Step 2. Navashesh of right side.
Cross out 9.427112=4+2+1+1=8

Step 3. Navashesh of left hand side = Navashesh of right hand side. The answer is verified.

Example 3: Verify the following multiplication by using Navashesh method
831x42x143x152=758629872

Step 1. Navashesh of left side.
Cross out 9 if any.
831=3
42=4+2=6
143=1+4+3=8
152=1+5+2=8
Therefore, 3x6x8x8=1152=1+1+5+2=9

Step 2. Navashesh of right side.
Cross out

9.758629872=5+8+6+8=27=2+7=9

Step 3. Navashesh of left hand side = Navashesh of right hand side. The correctness of the answer is verified.

Example 4: Verify the following result by Navashesh method.

159x-38=-6042

Step 1. Navashesh of left side.
Cross out 9 if any.
159=1+5=6
-38=-(3+8)=-11=-(1+1)=-2
Therefore, 6x-2=-12=-(1+2)=-3

Step 2. Navashesh of right side.
Cross out
9.-6042=-(6+0+4+2)=-12=-(1+2)=-3

Step 3. Navashesh of left hand side = Navashesh of right hand side. The result is verified.

Example 5: Verify the following multiplication by Navashesh method.

152x-31x-12x-59=-3336096

Step 1. Navashesh of left side.
Cross out 9 if any.
152=1+5+2=8
-31=-(3+1)=-4
-12=-(1+2)=-3
-59=-5
Therefore,
8x-4x-3x-5=-480=-(4+8+0)=-12=-(1+2)=-3

Step 2. Navashesh of right side.
Cross out
9.-3336096=-(6+0+6)=-12=-(1+2)=-3

Step 3. Navashesh of left hand side = Navashesh of right hand side. The answer is correct.

How to gain speed

The key to speed in Navashesh method is addition. The above discussion has shown the efficacy of Navashesh method in verifying results obtained in addition, subtraction and multiplication. The addition of all digits mentally, crossing out 9 and crossing out digits adding to 9 is the fast approach.

Navashesh method as applied to division

In division we all know the relation:

Divisor x Quotient + Remainder = Dividend.

Thus the Navashesh of divisor x Navashesh of quotient + Navashesh of remainder = Navashesh of dividend. The above rule will be clear from few examples.

Example 1: Divide 37637 by 7 and check the answer with Navashesh method.

Step 1.

```
7) 37637 (5376
   35
    26
    21
     53
     49
      47
      42
       5
```

Step 2. Navashesh of divisor 7=7

Step 3. Navashesh of quotient 5376=5+7=12=1+2=3

Step 4. Navashesh of remainder 5=5

Step 5. Applying the Navashesh formula for division

Therefore, Navashesh of left hand side 7x3+5=26=2+6=8

Step 6. Navashesh of right hand side.

Navashesh of dividend, cross out 9
37637=17=1+7=8

Thus the result of above division is verified.

Example 2: Divide 23923 by 9, and check the answer with Navashesh method.

Step 1.

```
9) 23923 (2658
   18
    59
    54
     52
     45
      73
      72
       1
```

Step 2. Navashesh of divisor 9=9

Step 3. Navashesh of quotient
2658=2+6+5+8=21=2+1=3

Step 4. Navashesh of remainder 1=1

Step 5. Applying the Navashesh formula for division i.e. Divisor x Quotient + Remainder = Dividend. Therefore, Navashesh of left hand side 9x3+1=28=2+8=10=1+0=1

Step 6. Navashesh of right hand side.
Navashesh of dividend, cross out 9
23923=2+3+2+3=10=1+0=1

Thus the result of above the division is verified.

Example 3: Divide 76854 by 8 and verify the result so obtained by Navashesh method.

Step 1.

```
8) 76854 (9606
   72
    48
    48
      54
      48
       6
```

Step 2. Navashesh of divisor 8=8

Step 3. Navashesh of quotient, cross out 9
9606=6+0+6=12=1+2=3

Step 4. Navashesh of remainder 6=6

Step 5. Applying the Navashesh formula for division
i.e. D x Q+R=D
Therefore, Navashesh of left hand side
8x3+6=30=3+0=3

Step 6. Navashesh of right hand side.
Navashesh of dividend, cross out 9
76854=7+6+8=21=2+1=3
Since Navashesh of left hand side = Navashesh of right hand side, the result of above division is verified.

Example 4: Divide 34892 by 12 and verify the answer thus obtained by Navashesh method.

Step 1.
```
12) 34892 (2907
    24
    108
    108
      92
      84
       8
```

Step 2. Navashesh of divisor 12=1+2=3

Step 3. Navashesh of quotient, cross out 9.2907=0

Step 4. Navashesh of remainder 8=8

Step 5. Applying the Navashesh formula for division
i.e. D x Q+R=D
Therefore, Navashesh of left hand side
3x0+8=8

Step 6. Navashesh of right hand side.
Navashesh of dividend, cross out 9
34892=8

As Navashesh of left hand side = Navashesh of right hand side, the result is verified.

Exercise 6

1. Carry out addition and check the answer by Navashesh method.

i) 846123+728321 ii) 363239+177190

iii) 723068+91129 iv) 462142+806

2. Carry out subtraction and check the answer by Navashesh method in the following.

i) 723068-91129 ii) 217829-9183

iii) 432157-81623

3. Carry out multiplication and check the answers.

i) 23x23 ii) 91x25 iii) 364x623

iv) 203x133 v) 312x212 vi) 84x67

4. Carry out division and check the answers.

i) 3) 649 ii) 8) 544

iii) 3) 2991 iv) 4) 927

Simple Equations

Several special types of equations can be solved practically on sight, with the aid of a beautiful Sutra, which reads: 'Sunyam Samysamuccaye' which merely says: When the Samuccaya is the same, that Samuccaya is zero i.e. it should be equated to zero.

Samuccaya is a technical term and has several meanings in different contexts.

First meaning

The first meaning of Samuccaya means a term which occurs as a common factor in all the terms concerned. For example in the equation: $12x+6x=18x+5x$

x occurs as common factor and therefore $x=0$. Similarly in equation: $8(x+1)+7(x+1)=3(x+1)$

$x+1$ is a common factor and therefore $x+1=0$ or $x=-1$

Second meaning

The second meaning of the word Samuccaya implies the product of the independent terms. Thus in equation: $(x+8)(x+9)=(x+6)(x+12)$

The product of independent terms on the left hand side $8x9=72$ is equal to the product of independent terms on the right hand side i.e. $6x12=72$. Therefore $x=0$. Normal method of solving the above equation will involve the following steps.

$x^2+9x+8x+72=x^2+12x+6x+72$

$x^2+17x+72=x^2+18x+72$

$x=0$

Third meaning

Samuccya thirdly means the sum of the denominators of two fractions having the same numerical numerator. This is clear from following examples.

Example 1. Solve $\frac{1}{2x-1} + \frac{1}{3x-1} = 0$

Steps The sum of denominator in the above equation

$2x-1+3x-1=5x-2$

Equating this to zero, we get

$5x-2=0$

$x=5=2/5$

Example 2. Solve $\frac{1}{3x-2} + \frac{1}{4x-6} = 0$

Steps The sum of denominator

$3x-2+4x-6=7x-8$

Equating this to zero, we get

$7x-8=0$

$x=8/7$

Fourth meaning

In this Samuccaya means total. It has several meanings in different contexts. These are explained below.

i) If the sum of the numerator and the sum of the denominator be same then that is zero. Thus $\frac{2x+9}{2x+7} = \frac{2x+7}{2x+9}$

Sum of numerator $N1+N2=2x+9+2x+7$

$=4x+16$

Sum of denominator $D1+D2=2x+7+2x+9x$

$=4x+16$

Equating this to zero we get

$4x+16=0$

$x=-4$

Compare and contrast this with the conventional method.

$$\frac{2x+9}{2x+7}=\frac{2x+7}{2x+9}$$

Cross multiplication will give

$$(2x+9)^2=(2x=7)^2$$
$$4x^2+36x+81=4x^2+28x+49$$
$$36x-28x=49-81$$
$$8x \quad =-32$$
$$x=-4$$

In fact as soon this characteristic i.e. N1+N2=D1+D2 of an equation is noticed the answer can be mentally found.

ii) If in the numerical total there is a common factor that should be removed. For instance

$$\frac{3x+4}{6x+7}=\frac{x+1}{2x+3}$$

N1+N2=3x+4+x+1=4x+5
D1+D2=6x+7+2x+3=8x+10
In this 2 is common. Thus D1+D2=2(4x+5). Here 4x+5 is there in both the sides and solution can be written.
4x+5=0
x=-5/4

Fifth meaning
And application for quadratics: Consider the following equation

$$\frac{3x+4}{6x+7}=\frac{5x+6}{2x+3}$$

As co-efficients of x^2 on the left-hand side and right hand side obtained after cross multiplication are 6 and 30 respectively the x^2 terms do not cancel.
(3x+4)(2x+3)=(6x+7)(5x+6)
$6x^2+17x+12=30x^2+71x+42$
$24x^2+54x+30=0$ is a quadratic equation. However, we notice in the equation $\frac{3x+4}{6x+7}=\frac{5x+6}{2x+3}$

i) N1+N2=3x+45x+6=8x+10
D1+D2=6x+7+2x+3=8x+10.
As N1+N2=D1+D2 we can straight away write
8x+10=0 or x=-5/4 as one solution.

ii) Let us now calculate
N1~D1=3x+4-6x-7
=(3x+3)
N2~D2=5x+6-2x-3
=(3x+3)

As they are equal we can write the other root of the quadratic equation 3x+3=0 or x=-1

Sixth meaning

This deals with equations of the type

$$\frac{1}{x-7} + \frac{1}{x-9} = \frac{1}{x-6} + \frac{1}{x-10}$$

The conventional method will involve taking the LCM and the solution will involve the following steps:

$$\frac{1}{x-7} + \frac{1}{x-9} = \frac{1}{x-6} + \frac{1}{x-10}$$

$$\frac{x-9+x-7}{(x-7)(x-9)} = \frac{x-10+x-6}{(x-6)(x-10)}$$

$$\frac{2x-16}{(x-7)(x-9)} = \frac{2x-16}{(x-6)(x-10)}$$

Cross multiplication

(2x-16)(x-6)(x-10)=(2x-16)(x-7)(x-9)

As 2x-16 is common we can write

2x-16=0
x=8

Instead of the above steps the vedic sutra simply tells us if D1+D2=D3+D4 then put that equal to zero and that is all.

In the above case D1+D2=x-7+x-9
=2x-16
and D3+D4=x-6+x-10
=2x-16

Since D1+D2=D3+D4

The solution is obtained by following 2x-16=0 or x=8

Sometimes we may find an equation where D1+D2=D3+D4 may not be apparent at the first instance. Some modification will be required to be done. Such cases are dealt with below.

Type 1: We may have an equation of the type

$$\frac{1}{x-8} - \frac{1}{x-5} = \frac{1}{x-12} - \frac{1}{x-9}$$

In such a case we transfer the –ve terms and re-write the equation

$$\frac{1}{x-8} + \frac{1}{x-9} = \frac{1}{x-12} + \frac{1}{x-5}$$

Now D1+D2=x-8+x-9=2x-17

and D3+D4=x-12+x-5=2x-17

as they are equal the solution can be written as

2x-17=0 or x=17/2

Type 2: Here we deal with equations of the type

$$\frac{x-2}{x-3} + \frac{x-3}{x-4} = \frac{x-1}{x-2} + \frac{x-4}{x-5}$$

Here again D1+D2=2x-7 and

D3+D4=2x-7 but numerator are not in proper form.

Therefore we re-write the above equation as

$$\frac{x-3+1}{x-3} + \frac{x-4+1}{x-4} = \frac{x-2+1}{x-2} + \frac{x-5+1}{x-5}$$

$$1+\frac{1}{x-3} +1+ \frac{1}{x-4} = \frac{1}{x-2} +1+1+ \frac{1}{x-5}$$

$$\frac{1}{x-3} + \frac{1}{x-4} = \frac{1}{x-2} + \frac{1}{x-5}$$

Now we can apply D1+D2=D3+D4

and the solution is 2x-7=0

x=3/2

Type 3: This deals with the equation of the type

Example 1. $\frac{2}{2x+3} + \frac{3}{3x+2} = \frac{1}{x+1} + \frac{6}{6x+7}$

Steps At first sight it appears to be different from the equations we have been dealing with. However, a slight modification will ensure that the above equation is not different.

$$\frac{3*2}{3(2x+3)} + \frac{2*3}{2(3x+2)} = \frac{6}{6(x+1)} + \frac{6}{6x+7}$$

$$\frac{6}{6x+9} + \frac{6}{6x+4} = \frac{6}{6x+6} + \frac{6}{6x+7}$$

Canceling 6 from the numerator

we get $\frac{1}{6x+9} + \frac{1}{6x+4} = \frac{1}{6x+6} + \frac{1}{6x+7}$

D1+D2=12x+13

D3+D4=12x+13

Therefore the solution is 12x+13=0 or x=-13/12

Consider another equation of similar type

Example 2. $\frac{3}{3x+1} + \frac{2}{2x-1} = \frac{3}{3x-2} + \frac{2}{2x+1}$

Steps Modifying the above equation we can write

$$\frac{2*3}{2(3x+1)} + \frac{3*2}{3(2x-1)} = \frac{2*3}{2(3x-2)} + \frac{3*2}{3(2x+1)}$$

$$\frac{6}{6x+2} + \frac{6}{6x-3} = \frac{6}{6x-4} + \frac{6}{6x+3}$$

Canceling 6 from the numerator we get

D1+D2=12x-1

D3+D4=12x-1

Therefore the solution is 12x-1=0 or x=1/12

Example 3. Solve the following equation

$$\frac{2x}{x+5} - \frac{9x-9}{3x-4} = \frac{4x+13}{x+3} - \frac{15x-47}{3x-10}$$

Steps Write down the co-efficient of x in the numerator and denominator and we get

2/1 - 9/3=4/1 - 15/3. This test can be applied in all the above cases and we are satisfied to proceed further.

Dividing the numerator in the above equation we can re-write the above equation as

$$2+\frac{1}{x+5}-\left(3+\frac{3}{3x-4}\right)=4+\frac{1}{x+3}-5-\frac{3}{3x-10}$$

$$-1+\frac{1}{x+5}-\frac{3}{3x-4}=-1+\frac{1}{x+3}-\frac{3}{3x-10}$$

$$\frac{1}{x+5}+\frac{3}{3x-10}=\frac{1}{x+3}+\frac{3}{3x-4}$$

$$\frac{3}{3x+15}+\frac{3}{3x-10}=\frac{3}{3x+9}+\frac{3}{3x-4}$$

D1+D2=6x+5

D3+D4=6x+5

Therefore, the solution is 6x+5=0 or x=-5/6

Example 4. Solve the following equation

$$\frac{5-6x}{3x-1}+\frac{2x+7}{x+3}=\frac{31-12x}{3x-7}+\frac{4x+21}{x+5}$$

Steps Write down the co-efficient of x in the numerator and the denominator and we get -6/3 + 2/1=-12/3 + 4/1 and as this is true we can apply this method.

Dividing the numerator in the above equation we can re-write the above equation as

$$-2+\frac{3}{3x-1}+2+\frac{1}{x+3}=-4+\frac{3}{3x-7}+4+\frac{1}{x+5}$$

$$\frac{3}{3x-1}-\frac{1}{x+4}=\frac{3}{3x-7}+\frac{1}{3x+5}$$

$$\frac{3}{3x-1}+\frac{3}{3x+9}=\frac{3}{3x-7}+\frac{3}{3x+15}$$

D1+D2=3x-1+3x+9=6x+8

D3+D4=3x-7+3x+15=6x+8

As they are equal 6x+8=0 or x=-8/6=-4/3

Exercise 7

Solve for x.

i) $\frac{1}{x+7} + \frac{1}{x+9} = \frac{1}{x+6} + \frac{1}{x+10}$

ii) $\frac{1}{x-7} + \frac{1}{x+9} = \frac{1}{x+11} + \frac{1}{x-9}$

iii) $\frac{1}{x-8} + \frac{1}{x-9} = \frac{1}{x-5} + \frac{1}{x-12}$

iv) $\frac{1}{x+1} + \frac{1}{x+3} = \frac{1}{x+2} + \frac{1}{x+4}$

v) $\frac{1}{x+1} + \frac{1}{x-3} = \frac{1}{x-4} + \frac{1}{x-8}$

Solving Simultaneous Simple Equations

The conventional cross multiplication of solving simultaneous equation method states that one of the variables is eliminated is fairly satisfactory. However, it involves quite a fair amount of effort.

The vedic method using Paravartya Sutra enables us to find the answer immediately by mere mental calculations.

Consider the following simultaneous equations:

3x + 4y = 6———equation 1

4x + 5y =14———equation 2

In the above equations we find the value of either x or y. Once the value of any of the variable is known the other can be found by substitution. Let us find the value of x.

Step 1. x=Numerator/Denominator=N/D

Step 2. Numerator = (co-efficient of x in the first row x constant in the second row)-(co-efficient of y in the second row x constant in the first row).

Thus the new numerator (N)=4x14-6x5
=56-30
=26

Step 3. Denominator = (co-efficient of y in the first row x co-efficient of x in the second row)-(co-efficient of y in the second rowx co-efficient of x in the first row).

Thus we go from the upper row across to lower one i.e. towards the left. Thus the denominator (D)=4x4-5x3

=16-15

=1

Step 4. Having found out the numerator and the denominator we can write x=26/1=26

Step 5. Substitute x=26 in above equation 1 or 2 to get value of y.

Thus y=-18.

Consider a few more examples.

Example 1: Solve the following equation for x and y

x-y=7

5x+2y=42

Step 1. x= N/D

Step 2. N=(-1)(42)-(7)(2)

=-42-14

=-56

Step 3. D=(-1)(5)-(2)(1)

=-5-2

=-7

Step 4. x=-56/-7=8

Step 5. Substitute x=8 in the above equation 1 or 2 to get value of y.

Thus 8-y=7 therefore y=1.

Example 2: Solve the following equation for x and y

2x+y=5

3x-4y=2

Step 1. x= N/D

Step 2. N=(1)(2)-(-4)(5)

=2+20

=22

Step 3. D=(1)(3)-(-4)(2)

=3+8

=11

Step 4. $x=22/11=2$

Step 5. Substitute $x=2$ in the above equation 1 or 2 to get the value of y.
Thus $2x2+y=5$ therefore $y=1$.

Example 3: Solve following equation for x and y
$11x+6y=28$
$7x-4y=10$

Step 1. $x= N/D$

Step 2. $N=(6)(10)-(-4)(28)$
$=60+112$
$=172$

Step 3. $D=(6)(7)-(-4)(11)$
$=42+44$
$=86$

Step 4. $x=172/86=2$

Step 5. Substitute $x=2$ in the above equation 1 or 2 to get value of y.
Thus $11x2+6y=28$
$6y=28-22=6$ therefore $y=1$.

First Special type of simultaneous equations
Some simple equations involving large co-efficient may appear as difficult but they can be readily solved using Anurupya Sunyam Anyat Sutra. This Sutra states that if one is in ratio the other one is zero.
Application of above Sutra will be clear from following examples.

Example 1: Solve for x and y from the following equations
$12x+8y=7$
$16x+16y=14$

Step 1. Here co-efficients of y are in the same ratio as independent terms i.e. 8:16::7:14
Therefore $x=0$

Step 2. Substituting $x=0$ in one of the above equation we get the value of y.
$8y=7$ therefore $y=7/8$

Example 2: Solve for x and y from the following equations
$12x+78y=12$
$16x+16y=16$

Step 1. Here co-efficient of y are in same ratio as independent terms i.e. 12:16::12:16
Therefore $y=0$

Step 2. Substituting $y=0$ in one of the above equation we get the value of x.
$12x=12$ therefore $x=1$

Example 3: Solve for x and y from the following equations
$499x+172y=212$
$9779x+387y=477$

Step 1. Here $172=(4)(43)$ and $387=(9)(43)$
$212=(4)(53)$ and $477=(9)(53)$
Thus the ratio between co-efficient of y and independent terms are the same
$172/387=(4)(43)/(9)(43)=4/9$
$212/477=(4)(53)/(9)(53)=4/9$
Therefore $x=0$

Step 2. Substituting $x=0$ in one of the above equations we get the value of y.
$172y=212$. Therefore, $y=212/172=(4)(53)/(4)(43)=53/43$
$y=53/43$.

Second Special Type of Simultaneous equation
There is another special type of simultaneous equation in which the x co-efficient and y co-efficient are found interchanged. In such cases we make use of Sutra

Saikalana-Vyavakalanabhyam which implies "by addition and subtraction".

Application of above Sutra will be clear from following examples.

Example 1: Consider the following simultaneous equations

45x-23y=113——equation 1

23x-45y=91———equation 2

In the above equations the co-efficient of x and y are interchanged therefore we can apply the method mentioned above.

Step 1. By addition of equation 1 and 2 we get

68x-68y=204

68(x-y)=204

x-y=204/68=3———equation 3

Step 2. By subtracting equation 1 and 2 we get

22x+22y=22

22(x+y)=22

x+y=22/22=1———equation 4

Step 3. Now we get the new set of equations

x-y=3

x+y=1

Step 4. Now these can be readily solved giving x=2 and y=-1

Example 2: Consider the following simultaneous equations

12x-8y=110——equation 1

8x-12y=70———equation 2

In the above equations co-efficients of x and y are interchanged therefore above method can be adopted.

Step 1. By addition of equation 1 and 2 we get

20x-20y=180

20(x-y)=180

$x-y=180/20=9$———equation 3

Step 2. By subtracting equation 1 and 2 we get

$4x+4y=40$

$4(x+y)=40$

$x+y=40/4=10$———equation 4

Step 3. Now we get a new set of equations

$x-y=9$

$x+y=10$

Step 4. Now this can be readily solved and we get

$x=19/2$ and $y=1/2$

Example 3: Consider the following simultaneous equations

$12x+17y=53$——equation 1

$17x+12y=63$———equation 2

In the above equations co-efficients of x and y are interchanged therefore the above method can be applied.

Step 1. By addition of equation 1 and 2 we get

$29x+29y=116$

$29(x+y)=116$

$x+y=116/29=4$———equation 3

Step 2. By subtracting equation 1 and 2 we get

$-5x+5y=-10$

$5(-x+y)=-10$

$x-y=-10/-5=2$———equation 4

Step 3. Now we get a new set of equations

$x+y=4$

$x-y=2$

Step 4. Now this can be readily solved and we get

$x=3$ and $y=1$

Exercise 8

1. Solve for x and y

i) $2x+y=5$ $3x-4y=2$	ii) $5x-3y=11$ $6x-5y=9$	iii) $x-y=7$ $5x+2y=42$
iv) $12x+7y=26$ $8x-4y=10$	v) $12x+8y=7$ $16x+16y=14$	vi) $12x+78y=12$ $16x+16y=16$
vii) $6x+7y=8$ $19x+14y=16$	viii) $45x-23y=113$ $23x-45y=91$	ix) $37x+29y=95$ $29x+37y=103$

Co-ordinate Geometry: Straight Lines

Straight lines constitute a very important chapter of co-ordinate Geometry. Conventional methods of solving various problems are adequate. Some of the problems, however, can be easily solved with vedic methods. These are being dealt with in this chapter.

Equation of a straight line passing through two given points:

Through a point many lines pass but through two given points there is only one unique line.

Let us find the equation of a straight line passing through two points A and B. The co-ordinates of points are:

A $(x_1 y_1)$=9,17

B $(x_2 y_2)$=7, -2

Conventional method 1:

Let us assume that the general equation of straight line y=mx+c passes through points A and B. In order to find variables 'm' and 'c' we substitute the co-ordinates of the points A and B.

Thus we get the equation

9m+c=17

7m+c=-2

Solving for 'm' and 'c' we get the desired equation.

y=19/2x-137/2 or 2y=19x-137

Conventional method 2:

The second method uses the formula

$y-y_1=y_2-y_1/x_2-x_1(x-x_1)$. Let us substitute values of x_1, y_1 and x_2, y_2.

We get $y-17=-2-17/7-9\ (x-9)$

$y-17=19/2(x-9)$

$2y-34=19x-171$

$2y=19x-137$

Vedic method

The vedic method makes use of the Paravartya Sutra. This enables us to write down the answer mechanically down by a mere casual look, at the given co-ordinates. The Sutra tells us to take following steps and write the equation as ax-by=c

Step 1: Calculation of 'a'

Write down the difference of 'y' co-ordinates as the x co efficient. Thus our x co-efficient in the above problem is

$y_1-y_2=17-(-2)=19=a$

Step 2: Calculation of 'b'

Write down the difference of 'x' co-ordinates as the y co-efficient. Thus our y co-efficient in the above problem is

$x_1-x_2=9-7=2=b$

Step 3: Calculation of 'c'

We got the equation 19x-2y=c where c is yet to be found. Since the line passes through (9, 17). We substitute and find

19*9-2*17=c

171-34=c

137=c

Thus our equation is

19x-2y=137

Let us consider a few more examples.

Example 1. Find the equation of a straight line passing through

$A\ (x_1y_1)=(9, 7)$ and

$B\ (x_2y_2)=(-7, 2)$

Step 1. Let the equation be ax-by=c

Step 2. $a=y_1-y_2=7-2=5$

Step 3. $b=x_1-x_2=9-(-7)=16$

Step 4. Equation is 5x-16y=c it passes through (9, 7) therefore

5*9-16*7=c

45-112=c

-67=c

Step 5. The equation is 5x-16y=-67

Example 2. Find the equation of a straight line passing through the points

$A\ (x_1y_1)=(10, 5)$ and

$B\ (x_2y_2)=(18, 9)$

Step 1. Let the equation be ax-by=c

Step 2. $a=y_1-y_2=5-9=-4$

Step 3. $b=x_1-x_2=10-18=-8$

Step 4. The equation is –4x+8y=c and it passes through (10, 5), therefore

-4*10+8*5=c

-40+40=c

0=c

Step 5. The equation is -4x+8y=0

-x+2y=0

x=2y

Example 3. Find the equation of a straight line passing through points

$A\ (x_1y_1)=(10, 8)$ and

$B\ (x_2y_2)=(9, 7)$

Step 1. Let the equation be ax-by=c

Step 2. $a=y_1-y_2=8-7=1$

Step 3. $b=x_1-x_2=10-9=1$

Step 4. The equation is $x-y=c$ it passes through (10, 8) therefore

$10-8=c$

$2=c$

Step 5. The equation is $x-y=2$

General equation representing two straight lines

When does a general equation of second order like, $ax^2+2hxy+by^2+2gx+2fy+c=0$ represent a part of straight lines? How does one find them?

Conventional method:

Let us take up the equation $12x^2+7xy-10y^2+13x+45y-35=0$ equation 1

The test for the above equation to represent a pair if straight lines is $abc+2fgh-af^2-bg^2-ch^2=0$ where various co-efficient are obtained from general equation $ax^2+2hxy+by^2+2gx+2fy+c=0$ equation 2. Comparing the co-efficient we get

$a=12$, $h=7/2$, $b=-10$, $g=13/2$, $f=45/2$ and $c=-35$

Substituting these in above equation 1 we get

$12*(-10)*(-35)+2*45/2*13/2*7/2-12(45/2)^2-(-10)(13/2)^2-(-35)(7/2)^2$

$=4200+\frac{4095}{4}-6075+\frac{1690}{4}+\frac{1715}{4}-1875+\frac{7500}{4}$

$=0$ Therefore the equation $12x^2+7xy-10y^2+13x+45y-35=0$ represents a pair of straight lines. Solving it for x we get

$x^2+x*\frac{7y+10}{12}+\frac{7y+13}{24}=\frac{10y^2-45y+35}{12}+\left(\frac{7y+13}{24}\right)^2$

$=\left(\frac{23y-43}{24}\right)^2$

$\left(x+\frac{7y+13}{24}\right)^2=\left(\frac{23y-43}{24}\right)^2$ ——— equation 1

$x+\frac{7y+13}{24}=\frac{23y-43}{24}$ ——— equation 2

From first equation we get

$x = \frac{23y-43}{24} - \frac{7y+13}{24}$

$x = \frac{23y-7y-43-13}{24} = \frac{16y-56}{24} = \frac{2y-7}{3}$

3x=2y-7, as one of the straight line.

The other one is obtained from second equation.

$x + \frac{7y+13}{24} = \frac{-23+43}{24}$

$x = \frac{-23-7y-13+43}{24}$

$x = -\frac{30y+30}{24} = -\frac{5y+5}{4}$

4x=-5y+5

We can now compare the above laborious process with the Vedic method.

Vedic method

Looking at the first three terms of the equation, i.e. $12x^2+7xy-10y^2$ we mentally make a note that (4x+5y) and (3x-2y) are the two factors. Also we note that the absolute term -35 must have 7 and –5 as the two factors. Now apply Urdhva Tiryak Sutra and multiply

3x-2y+7

4x+5y-5

$12x^2+(15-8)xy+(-15x+28x-10y^2)+(10y+35y)-35$

We get $12x^2+7xy-10y^2+13x+45y-35$ therefore this equation represents two straight lines (3x-2y+7) and (4x+5y-5). Thus the whole problem has been solved beautifully without taking recourse to lengthy calculation. Consider another example.

Example 1. Find whether the equation $3x^2-5xy-2y^2+5x+11y-12$ represents two straight lines and find the equation of those lines.

Vedic method: Looking at the first three terms $3x^2-5xy-2y^2$ we notice that (3x+y) and (x-2y) are the factors. Also the constant term –12 must have –4 and 3 as the two factors. We write these two equations as

$$3x+y-4$$
$$x-2y+3$$

Multiplication of these two by Urdhva Tiryak Sutra gives
$3x^2+(xy-6xy)+(-4x+9x-2y^2)+(3y+8y)-12$
$=3x^2-5xy-2y^2+5x+11y-12$

Thus the above equation represents a pair of straight lines whose equations are $3x+y-4$ and $x-2y+3$

Conventional method: The equation $3x^2-5xy-2y^2+5x+11y-12$ will represent a pair of straight lines if,
$abc+2fgh-af^2-bg^2-ch^2=0$
$a=3,\ b=-2,\ c=-12,\ f=11/2,\ g=5/2,\ h=-5/2$

Substituting these in above equation we get
$3*(-2)(-12)+2(11/2)(5/2)(-5/2)-3(11/2)^2+2(5/2)^2-(-12)(5/2)^2$
$=72-\frac{25*11}{4}-3*\frac{11^2}{4}+\frac{5^2}{2}+3*25$
$=72-\frac{275}{4}-\frac{363}{4}+\frac{25}{2}+75$
$=72-\frac{319}{2}+\frac{25}{2}+75$
$=\frac{144-319+25+150}{2}$
$=0$

Therefore the above equation represents a pair of straight lines. To find the equation of straight lines we solve it for x

$$x^2+\frac{5-5y}{3}x=\frac{2y^2-11y+12}{3}$$

$$x^2+\frac{5-5y}{3}x+\left(\frac{5-5y}{6}\right)^2=\frac{2y^2-11y+12}{3}+\left(\frac{5-5y}{6}\right)^2$$

$$\left(x+\frac{5-5y}{6}\right)^2=\frac{2y^2-11y12}{3}+\frac{25+25y^2-50y}{36}$$

$$\left(\frac{6x+5-5y}{6}\right)^2=\frac{24y^2-132y+144+25+25y^2-50y}{36}$$

$$=\frac{49y^2-182y+169}{36}$$

$$=\frac{(7y-13)^2}{6}$$

Therefore the first equation is

$6x+5-5y=7y-13$

$6x+18-12y=0$

$x-2y+3=0$

The second equation is

$6x+5-5y=-7y+13$

$6x+2y-8=0$

$3x+y-4=0$

It shows that the conventional method is laborious and time consuming.

Exercise 9

1. Find equation of straight line passing through

i) (10, 5) and (18, 9) ii) (17, 9) and (13, -8)

iii) (9, 7) and (4, -6) iv) (4, 7) and (3, 5)

v) (15, 16) and (9-3)

2. Prove that the equation $8x^2+10xy-3y^2-2x+4y-1=0$ represents a pair of straight lines and find out those lines.

3. Prove that the equation $12x^2-23xy+10y^2-24x+23y+12=0$ represents a pair of straight lines and find out those lines.

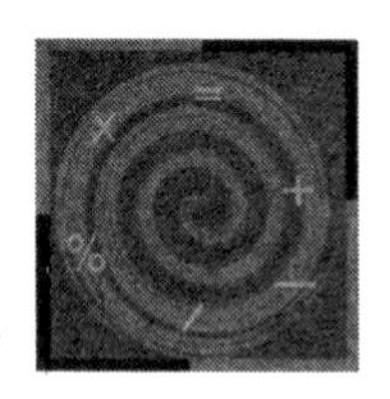

Answers

Exercise 1

1. 2744m^3 2. 5832 3. 97336

4. i) 17576 ii) 117649 iii) 238328
iv) 166375 v) 205379 vi) 103823
vii) 35937 viii) 24389 ix) 1021147343
x) 1092727 xi) 1015075125

Exercise 2

1. i) 14 ii) 23 iii) 34
iv) 54 v) 89 vi) 94

2. i) 14·08 ii) 23·74 iii) 34·11
iv) 54·09 v) 89·43 vi) 94·03

Exercise 3

1. i) $\cdot\dot{7}14285\dot{7}$ ii) $\cdot\dot{2}3076\dot{9}$ iii) ·2105263157899
iv) $\cdot\dot{5}7142\dot{8}$ v) $\cdot\dot{0}\dot{9}$ vi) $\cdot\dot{5}\dot{4}$
vii) $\cdot\dot{3}0769\dot{2}$ viii) ·0612244 ix) ·084745
x) ·1034482 xi) $\cdot\dot{1}2820\dot{5}$ xii) $\cdot\dot{2}8571\dot{4}$

Exercise 4

1. i) ·02564 ii) ·54676 iii) ·12034
iv) ·14775 v) ·05970 vi) ·09724
2. i) ·136 ii) ·126 iii) ·134
iv) ·070 v) ·124 vi) ·047
3. i) ·193633 ii) ·090146 iii) ·115577
iv) ·100858 v) ·051775 vi) ·092633
vii) ·244274 viii) ·099894
xi) ·123695

Exercise 5

1. i) 4 ii) 1 iii) 0 iv) 3
v) 6 vi) 4 vii) 8 viii) 6
ix) 3 x) 0

Exercise 6

1. i) 1574444
Navashesh of left hand side = Navashesh of right hand side=2
ii) 540429
Navashesh of left hand side = Navashesh of right hand side=6
iii) 814197
Navashesh of left hand side = Navashesh of right hand side=3
iv) 462948
Navashesh of left hand side = Navashesh of right hand side=6

2. i) 631939
Navashesh of left hand side = Navashesh of right hand side=4
ii) 208646
Navashesh of left hand side = Navashesh of right hand side=8

iii) 350534
Navashesh of left hand side = Navashesh of right hand side=2

3. i) 529
Navashesh of left hand side = Navashesh of right hand side=7
ii) 2275
Navashesh of left hand side = Navashesh of right hand side=7
iii) 226772
Navashesh of left hand side = Navashesh of right hand side=8
iv) 26999
Navashesh of left hand side = Navashesh of right hand side=8
v) 66144
Navashesh of left hand side = Navashesh of right hand side=3
ii) 5628
Navashesh of left hand side = Navashesh of right hand side=3

4. i) Q=216, R=1
Navashesh of left hand side = Navashesh of right hand side=1
ii) Q=68, R=0
Navashesh of left hand side = Navashesh of right hand side=4
iii) Q=997, R=0
Navashesh of left hand side = Navashesh of right hand side=3
iv) Q=231, R=3
Navashesh of left hand side = Navashesh of right hand side=3

Exercise 7

1. $x=-8$ 2. $x=-1$ 3. $x=17/2$ 4. $x=-5/2$
5. $x=7/2$

Exercise 8

1. i) $x=2, y=1$ ii) $x=4, y=3$ iii) $x=8, y=1$
iv) $x=87/52, y41/13$ v) $x=0, y=7/8$ vi) $x=1, y=0$
vii) $x=0, y=8/7$ ix) $x=2, y=-1$ x) $x=1, y=2$

Exercise 9

1. i) $2y=x$ ii) $4y-17x+253=0$ iii) $x+5y-44=0$
iv) $y-2x+1=0$ v) $6y-19x+189=0$
2. Equations of straight lines are
$4x-y+1=0$ and
$2x+3y-1=0$
3. The lines are
$4x-5y-4$ and
$3x-2y-3$